INDIA'S CHANGING INNOVATION SYSTEM

Achievements, Challenges, and Opportunities for Cooperation

Report of a Symposium

Committee on Comparative Innovation Policy:
Best Practice for the 21st Century

Board on Science, Technology, and Economic Policy

Policy and Global Affairs

Charles W. Wessner and Sujai J. Shivakumar, Editors

NATIONAL RESEARCH COUNCIL
OF THE NATIONAL ACADEMIES

THE NATIONAL ACADEMIES PRESS
Washington, D.C.
www.nap.edu

THE NATIONAL ACADEMIES PRESS   500 Fifth Street, N.W.   Washington, DC 20001

NOTICE: The project that is the subject of this report was approved by the Governing Board of the National Research Council, whose members are drawn from the councils of the National Academy of Sciences, the National Academy of Engineering, and the Institute of Medicine. The members of the committee responsible for the report were chosen for their special competences and with regard for appropriate balance.

This study was supported by: Contract/Grant No. SB1341-03-C-0032 between the National Academy of Sciences and the U.S. Department of Commerce; Contract/Grant No. OFED-381989 between the National Academy of Sciences and Sandia National Laboratories; and Contract/Grant No. NAVY-N00014-05-G-0288, DO #2, between the National Academy of Sciences and the Office of Naval Research. This material is based upon work also supported by the Defense Advanced Research Projects Agency Defense Sciences Office, DARPA Order No. K885/00, Program Title: Materials Research and Development Studies, Issued by DARPA/CMD under Contract #MDA972-01-D-0001. Additional funding was provided by Intel Corporation, International Business Machines, and Google. Any opinions, findings, conclusions, or recommendations expressed in this publication are those of the author(s) and do not necessarily reflect the views of the organizations or agencies that provided support for the project.

International Standard Book Number-13: 978-0-309-10483-8
International Standard Book Number-10: 0-309-10483-1

Limited copies are available from Board on Science, Technology, and Economic Policy, National Research Council, 500 Fifth Street, N.W., W547, Washington, DC 20001; 202-334-2200.

Additional copies of this report are available from the National Academies Press, 500 Fifth Street, N.W., Lockbox 285, Washington, DC 20055; (800) 624-6242 or (202) 334-3313 (in the Washington metropolitan area); Internet, http://www.nap.edu.

THE NATIONAL ACADEMIES
Advisers to the Nation on Science, Engineering, and Medicine

The **National Academy of Sciences** is a private, nonprofit, self-perpetuating society of distinguished scholars engaged in scientific and engineering research, dedicated to the furtherance of science and technology and to their use for the general welfare. Upon the authority of the charter granted to it by the Congress in 1863, the Academy has a mandate that requires it to advise the federal government on scientific and technical matters. Dr. Ralph J. Cicerone is president of the National Academy of Sciences.

The **National Academy of Engineering** was established in 1964, under the charter of the National Academy of Sciences, as a parallel organization of outstanding engineers. It is autonomous in its administration and in the selection of its members, sharing with the National Academy of Sciences the responsibility for advising the federal government. The National Academy of Engineering also sponsors engineering programs aimed at meeting national needs, encourages education and research, and recognizes the superior achievements of engineers. Dr. Charles M. Vest is president of the National Academy of Engineering.

The **Institute of Medicine** was established in 1970 by the National Academy of Sciences to secure the services of eminent members of appropriate professions in the examination of policy matters pertaining to the health of the public. The Institute acts under the responsibility given to the National Academy of Sciences by its congressional charter to be an adviser to the federal government and, upon its own initiative, to identify issues of medical care, research, and education. Dr. Harvey V. Fineberg is president of the Institute of Medicine.

The **National Research Council** was organized by the National Academy of Sciences in 1916 to associate the broad community of science and technology with the Academy's purposes of furthering knowledge and advising the federal government. Functioning in accordance with general policies determined by the Academy, the Council has become the principal operating agency of both the National Academy of Sciences and the National Academy of Engineering in providing services to the government, the public, and the scientific and engineering communities. The Council is administered jointly by both Academies and the Institute of Medicine. Dr. Ralph J. Cicerone and Dr. Charles M. Vest are chair and vice chair, respectively, of the National Research Council.

www.national-academies.org

Project Staff[*]

Charles W. Wessner
Study Director

Sujai J. Shivakumar
Senior Program Officer

McAlister T. Clabaugh
Program Associate

David E. Dierksheide
Program Officer

Paul Fowler
Senior Research Associate

Ken Jacobson
Consultant

Jeffrey C. McCullough
Program Associate

---

[*]As of December 2006.

For the National Research Council (NRC), this project was overseen by the Board on Science, Technology and Economic Policy (STEP), a standing board of the NRC established by the National Academies of Sciences and Engineering and the Institute of Medicine in 1991. The mandate of the STEP Board is to integrate understanding of scientific, technological, and economic elements in the formulation of national policies to promote the economic well-being of the United States. A distinctive characteristic of STEP's approach is its frequent interactions with public and private-sector decision makers. STEP bridges the disciplines of business management, engineering, economics, and the social sciences to bring diverse expertise to bear on pressing public policy questions. The members of the STEP Board[*] and the NRC staff are listed below:

Dale Jorgenson, *Chair*
Samuel W. Morris University Professor
Harvard University

Timothy Bresnahan
Landau Professor in Technology and
 the Economy
Stanford University

Lew Coleman
President
Dreamworks Animation

Kenneth Flamm
Dean Rusk Chair in International
 Affairs
Lyndon B. Johnson School of
 Public Affairs
University of Texas at Austin

Mary L. Good
Donaghey University Professor
Dean, Donaghey College of
 Information Science and Systems
 Engineering
University of Arkansas at Little Rock

Amo Houghton
Member of Congress, *retired*

David T. Morgenthaler
Founding Partner
Morgenthaler Ventures

Joseph Newhouse
John D. MacArthur Professor of
 Health Policy and Management
Harvard University

Edward E. Penhoet
President
Gordon and Betty Moore Foundation

Arati Prabhakar
General Partner
U.S. Venture Partners

William J. Raduchel
Independent Director and Investor

Jack Schuler
Chairman
Ventana Medical Systems

Suzanne Scotchmer
Professor of Economics and Public
 Policy
University of California at Berkeley

---

[*]As of December 2006.

STEP Staff*

Stephen A. Merrill
Executive Director

McAlister T. Clabaugh
Program Associate

David E. Dierksheide
Program Officer

Paul Fowler
Senior Research Associate

Charles W. Wessner
Program Director

Sujai J. Shivakumar
Senior Program Officer

Jeffrey C. McCullough
Program Associate

Mahendra Shunmoogam
Program Associate

---

*As of December 2006.

Contents

Preface

The United States faces a changing global environment where the capacity to innovate and commercialize new high-technology products is increasingly distributed worldwide. Governments around the world are taking active steps to renew and strengthen their national innovation systems, recognizing the strategic and economic importance of economic competitiveness.[1] In this new global environment, the United States must take up the challenge of maintaining its position of leadership by investing in its own capacity to innovate. The National Academies, in a recent report entitled *Rising Above the Gathering Storm*, called on the United States to adjust its policies concerning its workforce and research and development (R&D) capabilities to compete successfully in the future world economy.[2]

This report of a conference considers the opportunities, and some of the challenges of a strategic innovation partnership with India—a rising economic power

---

[1] National Research Council, *Innovation Policies for the 21st Century*, Charles W. Wessner, ed., Washington, D.C.: The National Academies Press, 2007.

[2] National Academy of Sciences/National Academy of Engineering/Institute of Medicine (NAS/NAE/IOM), *Rising Above the Gathering Storm: Energizing and Employing America for a Brighter Economic Future*, Washington, D.C.: The National Academies Press, 2007. The growing chorus of concern about U.S. innovation policy also included a report by the Council on Competitiveness, "Innovate America: Thriving in a World of Challenge and Change," Washington, D.C.: Council on Competitiveness, December 2004. Growing concerns about U.S. competitiveness led to the introduction in the Senate of the American Innovation and Competitiveness Act of 2006 and the Protect America's Competitive Edge Act of 2006. Also, in his 2006 State of the Union Address, President George W. Bush called for doubling commitment to basic research programs in physics and engineering over 10 years at the National Science Foundation (NSF), the Department of Energy (DOE), and the National Institute of Science and Technology (NIST) as a part of his Competitiveness Initiative. These initiatives have yet to become law, as this report goes to press.

BOX A
Innovation Ecosystem and Competitiveness

Innovation involves the transformation of an idea into a marketable product or service, a new or improved manufacturing or distribution process, or even a new method of providing a social service. This transformation involves an adaptive network of institutions that encompass a variety of informal and formal rules and procedures—a *national innovation ecosystem*—that shapes how individuals and corporate entities create knowledge and collaborate successfully to bring new products and services to market.

Competitiveness, in turn, refers to the ability of a nation's firms to produce the goods and services that can successfully compete in the globalized economic environment, while enabling a standard of living for its citizens that is both rising and sustainable. The ability of these factors to collaborate successfully depends on the flexibility and responsiveness of a nation's innovation ecosystem to recognize emerging opportunities and adapt to new challenges.

---

and an increasingly important locus of advanced research and development—in part through the growth of R&D facilities put in place by U.S. firms eager to draw on the intellectual assets and market opportunities of a rapidly growing India. The conference, held on June 17, 2006, at the National Academies in Washington, D.C., advances the joint communiqué following President Bush's state visit to India in March 2006, which called for strategic cooperation between the two nations in innovation and the development of advanced technologies.[3]

Cabinet ministers, senior officials, and academic experts from India and the United States came together at the conference on *India's Changing Innovation System* to explain the sources of India's exceptional recent economic performance, India's strengths in innovation, and the challenges India faces as it seeks to modernize its innovation system to become more competitive internationally as well as address the challenges of human development for its growing population. The conference, moreover, emphasized the opportunities that a strategic partnership in innovation holds for both the United States and India.

The conference, whose proceedings are reported in this volume, sought to highlight a set of complex and interrelated issues concerning India's changing innovation policies and the role the United States can play in aiding and benefiting from this transition. By necessity, even an ambitious one-day conference cannot (and did not) cover all facets of this rich topic. For example, the conference focused more on India's emerging strengths in the auto component manufacturing and pharmaceutical sectors than on the already familiar software and service sectors.

---

[3]The White House, "Fact Sheet: United States and India: Strategic Partnership," March 2, 2006 Press Release. For a broad overview of the evolution of the U.S.–India strategic partnership, see Teresita C. Schaffer, "Building a New Partnership with India," *Washington Quarterly*, 25(2):31–44, Spring 2002.

THE CONTEXT OF THIS REPORT

Since 1991 the STEP Board has undertaken a program of activities to improve policy makers' understanding of the interconnections among science, technology, and economic policy and their importance to the American economy and its international competitive position. The board's interest in comparative innovation policies derives directly from its mandate.

This mandate is reflected in STEP's earlier work on U.S. competitiveness, *U.S. Industry in 2000*, which assesses the determinants of competitive performance in a wide range of manufacturing and service industries, including those relating to information technology.[4] The Board also undertook a major study, chaired by Gordon Moore of Intel, on how government–industry partnerships can support the growth and commercialization of productivity-enhancing technologies.[5] Reflecting a growing recognition of the importance of the surge in productivity since 1995, the Board also launched a multifaceted assessment, exploring the sources of growth, measurement challenges, and the policy framework required to sustain the information and communications technology-based productivity gains and growth that have characterized the United States since the mid 1990s.[6]

Building on this experience, STEP's current study on Comparative Innovation Policy is developing a case-based international comparative analysis focused on U.S. and foreign innovation programs. The analysis includes a review of the goals, concept, structure, operation, funding levels, and evaluation of foreign programs similar to major U.S. programs, such as those found in Japan, Taiwan, Flanders in Belgium and now India. Among other activities, this study is convening a series of meetings with senior officials and academic analysts of these and other countries who are engaged in the operation and evaluation of these programs overseas, to gain a first-hand understanding of the goals, challenges, and accomplishments of these programs. As reflected in the conference reported in this volume, the National Academies Committee on Comparative Innovation Policy is also considering the role of innovation systems abroad and opportunities for collaboration that can complement the strengths of the U.S. innovation system in a globalizing innovation ecosystem.

---

[4]National Research Council, *U.S. Industry in 2000: Studies in Competitive Performance,* David C. Mowery, ed., Washington, D.C.: National Academy Press, 1999.

[5]This summary of a multivolume study provides the Moore Committee's analysis of best practices among key U.S. public–private partnerships. See National Research, *Government–Industry Partnerships for the Development of New Technologies: Summary Report,* Charles W. Wessner, ed., Washington, D.C.: The National Academies Press, 2003. For a list of U.S. partnership programs, see Christopher Coburn and Dan Berglund, *Partnerships: A Compendium of State and Federal Cooperative Programs,* Columbus, OH: Battelle Press, 1995.

[6]National Research Council, *Enhancing Productivity Growth in the Information Age: Measuring and Sustaining the New Economy,* Dale W. Jorgenson and Charles W. Wessner, eds., Washington, D.C.: The National Academies Press, 2006.

ACKNOWLEDGMENTS

We are grateful for the participation and the contributions of the Defense Advanced Research Projects Agency, the National Institute of Standards and Technology, the Office of Naval Research, and Sandia National Laboratories.

We are indebted to Ken Jacobson for his preparation of this meeting summary. Several members of the STEP staff also deserve recognition for their contributions, including McAlister Clabaugh, David Dierksheide, and Jeffrey McCullough for their role in organizing the conference and preparing this report for publication.

NATIONAL RESEARCH COUNCIL REVIEW

This report has been reviewed in draft form by individuals chosen for their diverse perspectives and technical expertise, in accordance with procedures approved by the National Academies' Report Review Committee. The purpose of this independent review is to provide candid and critical comments that will assist the institution in making its published report as sound as possible and to ensure that the report meets institutional standards for quality and objectivity. The review comments and draft manuscript remain confidential to protect the integrity of the process.

We wish to thank the following individuals for their review of this report: M.P. Chugh, Tata AutoComp Systems Ltd; Vinod Goel, The World Bank; Sarita Nagpal, Confederation of Indian Industry; Kesh Narayanan, National Science Foundation; and T.S.R. Subramanian, Government of India (Retired).

Although the reviewers listed above have provided many constructive comments and suggestions, they were not asked to endorse the content of the report, nor did they see the final draft before its release. Responsibility for the final content of this report rests entirely with the author and the institution.

William J. Spencer   Sujai J. Shivakumar   Charles W. Wessner

I

INTRODUCTION

India's Changing Innovation System
Achievements, Challenges, and Opportunities for Cooperation

India is a rising economic power and an increasingly important locus of innovation. Spurred by competition unleashed by a liberalization of once stifling regulations, India's private-sector firms are fast improving the quality of their products and services and are rapidly expanding their global presence. At the same time, U.S. and other multinational companies are increasingly locating their advanced research and development (R&D) operations in India to draw on the nation's highly trained scientists, engineers, and managers. In the process (and despite the endemic challenges of poverty) India is changing from a locus of low-cost contract research and reverse engineering to a global center of high-value, indigenously generated innovation. To sustain this transformation, Indian policy makers increasingly recognize the need for continuing economic reforms, new public investments in the nation's infrastructure, and new policy initiatives and institutions to encourage innovation, expand the skills and knowledge base of its population, and facilitate entrepreneurship.

As India grows as a center of global innovation, a new U.S.–India relationship is emerging—one where India is seen as both a partner and an effective competitor to the United States in the global marketplace. At the National Academies' June 2006 conference on India's Changing Innovation System, Ralph Cicerone, the president of the National Academy of Sciences, noted that advances in information and communications technology are creating new opportunities for the United States and India to benefit from the complementarities in their innovation systems.[1]

---

[1]See address by Under Secretary of State Nicholas Burns on "U.S. Policy in South Asia" to the Asia Society, on November 27, 2006, on the growing bilateral relationship with India. See also ad-

These technological advances encourage companies, individuals, and public institutions alike to adapt or be left behind. One of National Academies' own recent reports, addressed to the U.S. Congress, stressed that with the pace of global competition increasing, the United States must adjust its policies and institutions if it is to compete successfully in the future world economy.[2] And the United States is hardly alone in needing to adapt. Countries around the world, India included, are seeking to accelerate the transfer of scientific knowledge from universities, laboratories, and individuals into the marketplace. "In this process we must learn from each other," said Dr. Cicerone, calling this the "the entire premise" of the conference.

The conference, convened at the National Academies on June 17, 2006, examined many dimensions of the new U.S.–India innovation partnership. At the bilateral level, India and the United States have launched a new strategic relationship that specifically identifies science, technology, and innovation as a major focus of future relations.[3] This follows a variety of initiatives by the private sector in both countries to invest in promising firms, to make strategic acquisitions, and to draw on the distinct advantages of each other's innovation systems.

The conference also highlighted key developments within India, such as the dramatic improvements in the performance of India's national laboratories, and sampled some of the public debate ongoing in India concerning how the nation can expand its knowledge economy in a way that is socially inclusive as well as internationally competitive.

The conference also drew attention to opportunities for collaboration that can both help build India's innovation system and strengthen innovation in the United States. To maximize the positive potential of this relationship, India's Science and Technology (S&T) Minister, Kapil Sibal, emphasized the need for participants in India and the United States to learn from each other's policy initiatives and experiences. This introduction captures some of the key themes of this conference. The next section provides a detailed summary of the presentations and discussions of the National Academies' high-level conference on India's Changing Innovation System.

---

dress by David H. McCormick, Under Secretary of Commerce for Industry and Security at the World Economic Forum 2005 Summit on "India and the United States: An Emerging Global Partnership." He notes that changing geopolitics following the Cold War, the expansion of global markets, the information and communications technological revolution, and India's recent economic growth and economic openness are the foundations of a growing U.S.–Indian strategic partnership.

[2]NAS/NAE/IOM, *Rising Above the Gathering Storm: Energizing and Employing America for a Brighter Economic Future*, Washington, D.C.: The National Academies Press, 2007.

[3]The White House, "Fact Sheet: United States and India: Strategic Partnership," March 2, 2006 Press Release.

BOX A
Innovation Systems and Development

Popularized by Richard Nelson, the term *National Innovation System* (NIS) refers to a network of institutions in the public and private sectors, whose activities and interactions initiate, develop, modify, and commercialize new technologies.[a] The NIS concept highlights the linkages and knowledge flows among multiple and dispersed public and private organizations, including universities, research laboratories, and large and small businesses.[b]

Bringing innovative ideas to market involves complex interlinkages among industry, academia, and government within multiple overlapping "innovation ecosystems." This ecosystems approach emphasizes the importance of creating and improving institutions to interweave the different parts of a nation's innovation system.

In the context of a developing country such as India, innovation can provide a channel to both increase growth and reduce poverty. By applying knowledge in new ways to production processes, more, better, or previously unavailable products can be produced at prices that all Indians can afford. Public policies to enhance *pro-growth innovation* include improving higher education and creating new public–private partnerships as well as pursuing broad economic reforms that create the appropriate environment for investment in and commercialization of research. Bolstering *inclusive innovation* includes efforts to harness creative efforts for the poor; to promote, diffuse, and commercialize grassroots innovations; and to help the informal sector better absorb existing knowledge.[c]

---

[a]See Richard R. Nelson and Nathan Rosenberg, "Technical Innovation and National Systems," in *National Innovation Systems: A Comparative Analysis*, Richard R. Nelson, ed., Oxford: Oxford University Press, 1993, p. 4. Nelson notes that the idea of a "national innovation system" captures "a new spirit of what might be called 'techno-nationalism' . . . combining a strong belief that the technological capabilities of national firms are a key source of competitive prowess, with a belief that these capabilities are in a sense national, and can be built by national action." In Nelson, *National Innovation Systems*, op. cit., p. 5. The National Innovation System model appeals to policy makers because it provides an interpretive scheme that focuses on the nation as a unit of analysis. For a critique of the nation as a unit of analysis, see John de la Mothe and Gilles Paquet, "National Innovation Systems, 'Real Economies' and Instituted Processes," *Small Business Economics*, 11, 101–111, 1998.

[b]The emerging NIS literature draw attention to the presence of interactions and flows among public- and private-sector organizations in initiating, modifying, and diffusing new technologies. See P. Patel and K. Pavitt, "National Innovation Systems: Why They Are Important and How They Might Be Compared?" *Economic Change and Industrial Innovation*, 1994. See also C. Endquist (ed.) *Systems of Innovation: Technologies, Institutions, and Organizations*, London: Pinter, 1997.

[c]See World Bank, *Innovation Systems: World Bank Support of Science and Technology Development*, Vinod Kumar Goel, ed., Washington, D.C.: World Bank, 2004.

A NEW STRATEGIC PARTNERSHIP IN INNOVATION

Recognizing a need for deeper, more collaborative relationships among Indian and U.S. businesses, academic institutions, and governmental organizations, the joint communiqué following President Bush's state visit to India in March 2006 called for strategic cooperation in innovation and the development of advanced technologies.[4] This strengthened focus on the potential for collaboration between India and the United States to advance scientific research and support the process of innovation underscored the timeliness of the conference.

In her opening remarks at the conference, Paula Dobrainsky, the U.S. undersecretary for global affairs stressed that current collaborations between the United States and India in science, technology, and engineering-related research and development are genuine partnerships, not the assistance programs of the past. "India has the scientific and technological base to join the United States as equal partners in pushing forward the frontiers of research," she said, resulting in "a very positive impact on the lives on all of our citizens." She added that the key to mutual economic growth and prosperity lies in increasing linkages of the U.S. and Indian knowledge bases: the two nations' scientists, researchers, academics, and business leaders.

Together, these various agreements underscore the commitment of both nations to mutually beneficial cooperation. These linkages have been reinforced by the umbrella Science and Technology Framework Agreement signed in October 2005 by Secretary of State Rice and S&T Minister Sibal. The agreement establishes, for the first time, protocols to observe intellectual property rights, among other provisions necessary to facilitate active collaborative research across a broad range of disciplines. It builds the framework within which Indian and American scientists in government, the private sector, and universities can collaborate in basic and applied research in areas as diverse as information technology, agriculture, health, and energy.

Speaking at the conference, Secretary of Energy Samuel Bodman, who cochairs the U.S.–India Energy Dialogue, noted that India's participation in the Future-Gen international partnership (whose goal is to create a zero-emission coal-fired power plant) is an example of the type of cooperation that the United States hopes to encourage under this S&T framework. He also cited, as a potential turning point in bilateral relationship, the U.S.–India nuclear agreement, which foresees research cooperation in exploiting the potential of nuclear energy more safely.[5]

INDIA'S CHANGING ECONOMIC POLICIES

India's new dynamism, and hence these new opportunities for bilateral collaboration, are widely seen to have begun with the policies of economic

---

[4]For a broad overview of the evolution of the U.S.–India strategic partnership, see Teresita C. Schaffer, "Building a New Partnership with India," *Washington Quarterly*, 25(2):31–44, Spring 2002.

[5]The U.S.–India Peaceful Atomic Energy Cooperation Act was signed into law on December 18, 2006, by President George W. Bush, following overwhelming approval in both the House of Representatives and the Senate.

BOX B
A Partnership Driven by Science and Technology

In his remarks, India's ambassador to the United States, Ronen Sen, noted that science and technology is a major driver of the new U.S.–India Strategic Partnership. Current bilateral initiatives include:

- The **Next Steps in Strategic Partnership:** This 2005 bilateral agreement has extended the prospects for cooperation to civilian uses of nuclear, space, and dual-use technologies.
- A new, 10-year **Framework for the U.S.–India Defense Relationship:** Concluded in 2005 at the ministerial level, the framework has established a joint Defense Procurement and Production Group.
- A new **U.S.–India Bi-National Science and Technology Endowment Fund:** The purpose of this fund is to facilitate joint research projects with potential for industrial application.
- The **U.S.–India Knowledge Initiative on Agriculture:** A $100 million fund is bringing together research institutions and corporate entities in both countries to raise agricultural productivity and increase prospects for agroindustrial business in India.
- The **U.S.–India Energy Dialogue:** Launched in 2005, it envisages the **Agreement on Civil Nuclear Energy Cooperation** as well as cooperation in such areas as oil and gas, clean-coal technologies, and renewable energy sources.
- The **U.S.–India Working Group on Civil Space Cooperation:** This group is to renew and upgrade cooperation in space.
- The **Agreement on Civil Nuclear Energy Cooperation:** The agreement foresees collaboration to help develop more environmentally friendly and proliferation-resistant technologies.

---

liberalization started in the early 1990s and supported by successive national governments.

Montek Singh Ahluwalia, the deputy chairman of the Indian Planning Commission, and a key architect of India's economic reforms, noted that India's economic liberalization reflects the premise that the private sector is the critical driver of growth.[6] By promoting a competitive environment within India, the dynamic effects of the private sector have emerged, where foreign firms are increasingly allowed to participate in the Indian economy, and where private Indian firms are encouraged to compete internationally through new investments and acquisitions of offshore companies.

---

[6]Dani Rodrik and Arvind Subramanian argue that the acceleration in India's economic performance with the liberalization of 1991 was triggered by an attitudinal shift of the government toward a pro-business approach. See Dani Rodrik and Arvind Subramanian, "From 'Hindu Growth' to Productivity Surge: The Mystery of the Indian Growth Transition," NBER Working Paper 10376, 2004.

Policy Achievements and Targets

Mr. Ahluwalia also noted that India's economic reforms have deliberately been gradual and cautious.[7] In a pluralist democracy such as India's, he explained, a consensus for change has to be built across the social and political spectrum. Such a consensus, he averred, has taken root and is being reinforced by significant gains in the country's growth rate; despite these gains, the process of reform remains politically challenging.[8] This growth rate has consistently been above 7 percent over the previous 5 years, reaching 8.4 percent in 2005. India's Planning Commission, targeting continuing acceleration, hopes to push its growth rate to 9.5 percent per annum by the end of the following 5 years, while achieving an annual growth average of 8.5 percent over the entire period.

To realize this faster and more broadly based rate of growth, the government recognizes that it needs to improve agricultural productivity, aiming at a second Green Revolution through improved agroprocessing and modern marketing[9]; invest more in its educational system to provide Indians the marketable skills to integrate and compete in the world economy; build high-quality infrastructure to facilitate investment and promote trade; and focus on improving energy efficiency and security to fuel the growing economy.[10]

The Two Indias

While what has been accomplished in recent years is impressive, T. S. R. Subramanian reminded the audience that India's development challenges remain daunting.[11] Of India's nearly 1 billion inhabitants, an estimated 350–400 million are below the poverty line, with over three-fourths of the poor living in rural areas. Although the country has made great strides as an emerging knowledge

---

[7]For a sector-by-sector analysis of economic reforms in India, see Montek Singh Ahluwalia, "Economic Reforms in India Since 1991: Has Gradualism Worked?" *Journal of Economic Perspectives*, 16(2):67–88, 2002.

[8]Anand Giridharadas, "Growth Spurt in India Hides Government Gridlock," *International Herald Tribune*, Sept. 29, 2006.

[9]The second Green Revolution calls for policies to shift India's rural economy from a reliance on peasant farming to a new focus on agribusiness, thereby encouraging private capital to move from urban to rural areas. Proposed policies would lift distribution controls, allow large retailers to contract directly with farmers, invest in irrigation, and permit the consolidation of fragmented holdings. See Gurcharan Das, "The India Model," *Foreign Affairs*, 85(4), July/August 2006. The first Green Revolution in India refers to significant gains in agricultural productivity through the introduction after 1965 of new high-yield varieties of crops.

[10]These goals are reflected in the forthcoming 11th Five-year Plan of India's Planning Commission. See remarks by the Chairman of the Planning Commission, Prime Minister Manmohan Singh, on the October 18, 2006, on the release of the 11th Plan Approach Paper. Access at *<http://pmindia.nic. in/speech/content.asp?id=431>*.

[11]This point is also emphasized in the paper in this volume by Carl Dahlman, a leading World Bank analyst of the knowledge economy, now with Georgetown University.

economy, only 16 percent of the population has completed high school.[12] Real problems continue. Public health needs major improvements and rural infrastructure remains "abysmal."[13] The political consequences of the gulf between the two Indias are troubling, Mr. Subramanian noted. Yet even as modern information technologies increasingly inform the poor about disparities in wealth and opportunity, they also provide a medium for exposure and can be used effectively as a low-cost medium of education as well as encouragement to the poor to compete and strive in a free economy.[14]

Mr. Subramanian noted that while much of the reforms that Mr. Ahluwalia spoke of refer to India's federal government, most of what touches the average Indian—including law and order, education, public health, and rural development—falls under the purview of state governments according to India's Constitution. As a result, Mr. Subramanian said, the winds of change had not yet reached the poor in many places, not because of a lack of awareness but because of a lack of will arising from local political compulsions.[15] Patterns of rent seeking and corruption are entrenched in many state governments in India, inhibiting needed reforms.[16]

At the same time, some state governments have begun instituting their own reforms, working for example to upgrade and support the growth of their manu-

---

[12]Indeed, most Indians still live in a largely subsistence economy, with literacy rates of 73 percent among adult men and 48 percent among adult women, and with 35 percent of the population living below the international poverty benchmark of $1 a day. The World Bank estimates that only 4 percent of India's workforce is employed in the modern private sector, while 89 percent of the workforce is in the informal sector. Source: The World Bank, *World Development Indicators, 2006*. Access at <http://devdata.worldbank.org/wdi2006/contents/cover.htm>.

[13]Addressing this second challenge, the government of India in 2000 launched a $13.5 billion program of rural road construction that is to connect all habitations that have a population of more than 500 persons. About 160,000 rural habitations are to be covered under this program. Access at <http://www.pmgsy.nic.in/pmgsy.asp>.

[14]This note of caution tempering the enthusiasm about India's growth prospects has been echoed by India's leadership, including Prime Minster Manmohan Singh and Sonia Gandhi, the leader of the India's governing coalition, at the World Economic Forum's 2006 India summit. See *Associated Press*, "India Shows Confidence, Openness About Risks Confronting Economy," Nov. 28, 2006.

[15]For an examination of factors underpinning the variations in growth within India, see Montek Singh Ahluwalia, "State Level Reforms Under Economic Reforms in India," Working Paper No. 96, Stanford University, March 2001. For an additional analysis of the link between state and local policies (including restrictive labor policies) and the unequal effects of liberalization in India, see Phillipe Aghion, Robin Burgess, Stephen Redding, and Fabrizio Zilibotti, "The Unequal Effects of Liberalization: Theory and Evidence from India," Center for Economic Policy Research, March 2003. For an econometric analysis of income differentials at the state level following the 1991 liberalization, see Michelle Baddeley, Kirsty McNay, and Robert Cassen, 2006, "Divergence in India: Income Differentials at the State Level, 1970–97," *The Journal of Development Studies*, 42(6):1000–1022.

[16]For an incisive analysis of the structure of rent-seeking in some Indian states, see Robert Wade, "The Market for Public Office: Why the Indian State Is not Better at Development," *World Development*, 13:467–497, 1985.

facturing and services sectors.[17] Similar positive state-level commitments will be
necessary across the country for India to realize the second Green Revolution,
said Mr. Subramanian. Alleviating rural poverty will require new methods, management practices, and techniques of delivering technology assistance to farmers.
The private sector and voluntary agencies, Mr. Subramanian added, will have to
play a key role in this transformation.[18]

Strengthening India's Knowledge Economy

Acknowledging Mr. Subramanian's point that India presents an "extreme dual
economy," Carl Dahlman of Georgetown University noted that the country nevertheless possesses unique strengths and now faces new opportunities to leverage
these strengths to improve its competitiveness and the well-being of its people.[19]
India's fundamental strengths, he said, include its very large domestic market, young and growing population, critical mass of educated people, very strong
R&D infrastructure, and strong science and engineering capabilities centered in
areas such as chemicals, pharmaceuticals, and software. India is a world center
for many digital services, a location where "anything that can be offshored" can
be done very cost-effectively. From this base, India is becoming a center for innovation for multinational companies, which have already established around 400
R&D centers in India to draw on its scientists and engineers. The Indian diaspora,
strongly represented in the United States, also provides an excellent source of
everything from information and advice to access to markets, technology, and
financing as India's activities increase in sophistication.[20] India also benefits from

---

[17]The state of Tamil Nadu has, for example, been lauded by the World Bank as one of India's best-
performing states in recent years, based on its institutional reforms and changes in economic policy.
See *The Hindu*, "State Pins Hope on Growth Initiative," Aug. 4, 2005.

[18]Already, initiatives by major Indian corporations, including Reliance and Bharti, are beginning to
transform India's agricultural sector by developing the necessary supply-chain infrastructure linking
the farmer to new retail food supermarkets. See *Bloomberg News*, "The Next Green Revolution,"
Aug. 21, 2006. Innovations in supply-chain management are already helping to integrate more of
India's poor into the modern economy. For example, cyber kiosks called e-Choupals, set up in several
thousand villages, have transformed the way that farmers transact with industry by giving them power
of information, thus eliminating the middlemen. For a case study, see <*http://www.digitaldividend.
org/case/case_echoupal.htm*>. Also, Bharti Tele-Ventures offers some of the lowest phone prices in
the world, making telephone services affordable for the first time to many of India's poor. Based on
its innovative business strategy, Bharti has expanded its position as India's largest mobile service
provider. A third example, the development by Tata of a $2,000 car, was described by Mr. Chugh at
the conference.

[19]For an overview of potential growth trajectories and policy options to foster India's knowledge
economy, see Carl Dahlman, "India's Knowledge Economy in the Global Context," in this volume.
See also Carl Dahlman and Anuja Utz, *India and the Knowledge Economy: Leveraging Strengths and
Opportunities,* Washington, D.C.: World Bank Institute, 2005.

[20]For a discussion of polices for India to capitalize on its overseas diaspora, see Devesh Kapur,
"Indian Diaspora as a Strategic Asset," *Economic and Political Weekly*, 38(5):445–448.

relatively deep financial markets, far better than China's according to Dr. Dahlman, and its private businesses are beginning to strengthen their export orientation, seeking international strategic alliances and making acquisitions abroad.[21]

Needed Policy Measures

To build on these strengths, Dr. Dahlman outlined a series of complementary policy steps. To improve the economic regime, he recommended easing restrictions on firm entry and exit and the hiring and firing of workers. To improve the nation's infrastructure for primary and secondary education, he suggested greater use of private providers of education and training. To encourage greater innovation, he encouraged the development of partnerships between academia and industry. These partnerships, he noted, can also help increase universities' awareness of the skills required to create the knowledge workers necessary for economic progress. To close the digital divide, he recommended investments to broaden the penetration and use of innovative information and communications technologies. Finally, to improve the size and efficiency of the nation's R&D investments, he emphasized the need for policies that help attract more foreign direct investment, motivate greater private investment in R&D, and help entrepreneurs bring new products to the market.

Such entrepreneurship can already be seen in India's information technology, pharmaceutical, and automotive component industries, which are taking significant steps to improve manufacturing quality, increase the level of internal R&D, and develop global strategies for competition.

INDIA'S CHANGING INDUSTRY

Indian industry has made rapid progress since the economy was liberalized in 1991. In the days of the "License Raj," elaborate licenses, regulations, and the accompanying red tape were required to set up and conduct business in India. These regulations limited competition, providing Indian industry with little incentive to invest in research and development.[22] Many manufacturers engaged in contract manufacturing that required little indigenous innovation. The result was often stale products of poor quality for the domestic market. As R. A. Mashelkar, the president of the Indian National Science Academy, candidly observed at the conference, "there was no competitiveness because we were a closed economy. Industry produced gums that did not stick, yet people bought them. We produced

---

[21]Even so, some leading U.S. venture capital firms are exploring the possibility of taking companies public on Chinese stock exchanges. *International Herald Tribune*, "Start-ups Explore Abroad for IPOs," Dec. 25, 2006.

[22]See Rodrik and Subramanian, "From 'Hindu Growth' to Productivity Surge: The Mystery of the Indian Growth Transition," *op. cit.*

plugs that did not fit, and yet we bought them. We produced cars on which everything other than the horn made noise, and we bought them."

This state of affairs began to change with the initial deregulation of the economy. As a direct result, India's manufacturing sector is today increasingly dynamic and innovative. For India's government, supporting a renewed manufacturing sector is a national priority, not least for its potential to create jobs for India's youthful population.[23] This priority is reflected in India's National Manufacturing Competitiveness Council (NMCC) focus on increasing the contribution of manufacturing to India's GDP from 18 percent to 30 percent over the next 10 years. To this end, the NMCC has called for policy measures that improve the climate for innovation, including a technology development fund for industry, improvement of the science education system, as well as programs to promote the commercialization of advanced technology products.[24]

A Turnaround in Manufacturing Quality

This agenda for international competitiveness through enhancement of manufacturing innovation and quality is also being promulgated by the leaders of Indian industry. As Surinder Kapur, chairman of the Confederation of Indian Industry's (CII) Mission for Innovation in Manufacturing, noted at the conference, Indian industry recognizes that it cannot expect to survive in a competitive international environment based solely on its lower cost advantages. Indeed, the CII leadership recognizes that people-cost arbitrage that Indian companies have been relying on can keep it competitive in the international environment only in the short term. Although contract manufacturing is expected to provide continuing growth in the near future, Dr. Kapur affirmed that Indian companies need to reposition themselves as advanced manufacturers and service providers of creative products if they are to be competitive over the long term.

To help Indian industries achieve this transition to international competitiveness, CII—India's premier business association—has mounted repeated efforts to improve the quality and standards of Indian industry across the country. Dr. Kapur described a range of these initiatives, from sponsoring training in *process control* to imparting the tools of *continuous improvement*, to the introduction of *management by objectives*, which ensures the alignment of the entire firm toward its quality goals.

---

[23]The average Indian is now just 23 years old, with over half the population under the age of 25. *Financial Times*, "Engaging India: Demographic dividend or disaster?" Nov. 15, 2006.

[24]These policy initiatives were described by Mr. Krishnamurthy, the NMCC chairman, at the 2006 national conference on "Enhancing India's Manufacturing Competitiveness by Leveraging Indian R&D." *The Hindu*, "Stress on Innovation in Manufacturing," Nov. 17, 2006. For a review of the various NMCC initiatives, see NMCC, "The National Strategy for Manufacturing," New Delhi, March 2006.

Dr. Kapur noted that Indian industry must reinvent itself as ambidextrous business organizations that encourage innovative ideas and bring them to market.[25] One part of CII's strategy to promote this change is to target 100 promising Indian companies for intensive encouragement so that they can become leaders in innovation and product development. Acknowledging that this is a small number for a vast country, Dr. Kapur noted that the purpose of this initiative is to stimulate new ways of thinking by India's corporate leaders. By demonstrating what is possible, mindsets can be changed.[26] He said that CII is also seeking to encourage small and medium-sized businesses to gather in informal clusters so that, by pooling their resources, they can also learn how to bring quality-improving methods to production, distribution, and sale.

New Global Outreach

Indian industry has indeed come a long way in a short time. In addition to rapidly improving quality (documented by winning the Deming Prize and other top international prizes in manufacturing quality), several Indian firms, especially in the software and services sectors and more recently in its pharmaceutical and automotive component sectors, are now recognized as globally competitive, with active international partnering and acquisition strategies.[27] Corporate leaders at the conference on India's Changing Innovation System described new developments in the latter two sectors.

In the pharmaceutical sector, Dr. Swati Piramal noted that her company, Nicholas Piramal India Limited, is pursuing an ambitious strategy of growth through strategic acquisitions, alliances, and joint ventures that have brought it in partnership with many international firms including Allergan, Aventis, and Roche. Nicholas Piramal is also rapidly acquiring a global footprint by extending operations in the United Kingdom, the United States, Canada, and China. Already, she

---

[25]Recent empirical work supports the beneficial impacts of enhanced competition described by Dr. Kapur. See Satish Chand and Kunal Sen, "Trade Liberalization and Productivity Growth: Evidence from Indian Manufacturing," *Review of Development Economics*, 6:120, February 2002. The authors find that trade liberalization in Indian manufacturing has positively raised total factor productivity (TFP) growth.

[26]As evidence of the prevalence of advanced skills in India, Dr. Kapur posted a list of familiar names in U.S. research that are operating R&D centers in India: Bell Labs, Cognizant Technologies, Enercon, Exxon, GE Industrial Systems, GE Medical Systems, IBM, Intel, Lucent, Microsoft, Motorola, National Instruments, Oracle, SeaGate, Texas Instruments, and Xytel. "We believe there is a lot of culture that Indian manufacturing companies need to take advantage of," he observed.

[27]Tata's recent takeover of Corus Steel is the latest in a series of global acquisitions by Indian firms. Other Indian firms recently acquiring assets overseas include Bharat Forge, Ranbaxy, Wipro, and Nicholas Piramal. According to *The Economist*, Indian companies announced 115 foreign acquisitions, with a total value of $7.4 billion in the first three quarters of 2006. See *The Economist*, "India's Acquisition Spree," Oct. 12, 2006.

said, Nicholas Piramal employs about 1,000 workers at its three manufacturing facilities in the United States.

Dr. Piramal noted that her company's pipeline is rich, and its dream is to develop new drugs for the global market for $50 million—a relative bargain compared to the $1 billion cost of similar efforts in the United States. In India, she noted, Nicholas Piramal can buy "a lot of scientific horsepower" for the money. What is more, the company's new R&D investments appear to be bearing fruit; she reported that between January and June 2006 Nicholas Piramal filed 14 patents for New Chemical Entities.

Underscoring the new confidence of Indian industry, Dr. Piramal announced that on June 15, 2006, two days before the National Academies' conference, Nicholas Piramal had signed an agreement to acquire from Pfizer a 450-employee facility in Morpeth in the United Kingdom, providing Nicholas Piramal access to Pfizer's global sourcing network. This acquisition, she said, is consistent with Nicholas Piramal's intent to become a global leader in custom manufacturing across the pharmaceutical value chain.[28]

The Indian automotive component industry has similar global ambitions. M. P. Chugh of Tata Auto Component Systems (TACO) noted that Tata's vision of making a $2,000 car for the Indian mass market calls for the company to develop not only an innovative automobile but also an innovative business model to bring this vision to reality. This business model would "not only use the engineering talent in India, but leverage the engineering talent in India for a global business market."

When TACO was formed in 1996, the cutting-edge technologies required to produce advanced automotive components were not found in India. To overcome this hurdle, TACO has formed 16 global partnerships, including alliances with Johnson Controls and Visteon, to coordinate the efforts of engineers around the globe to conduct research and development around the clock. TACO, he reported, has also established 4 advanced engineering centers (including a center in the United States) and 16 manufacturing plants that produce components such as interior plastics, seating systems, exteriors and composites, and wiring harnesses. The key to the success of the Tata's joint ventures with U.S. companies such as Johnson Controls, he noted, lies in coordinating the efforts of engineers spread around the world as they carry on work on a given product development program.

TACO's aspirations are global. The company, Mr. Chugh said, believes that it needs to be not only in Asia, but also in the North American and European markets. He noted that while Chinese manufacturers are better at "shoot and ship"—that is, manufacturing a product given a drawing and design—Indian auto manufacturers aim to have the capacity to design, test, and validate as well as

---

[28]*Associated Press Financial Wire*, "Indian Drug Maker Nicholas Piramal in Deal to Acquire Pfizer's UK Facility," June 15, 2006.

manufacture automotive parts. Describing TACO's design capability as its core strength, Mr. Chugh noted that his company is committed to building a deep engineering and R&D base that would enable it to develop new technologies and innovative solutions for customers around the world.

Together, these three presentations by representatives of Indian industry underscore the global reach, current accomplishments, and aspirations of India's leading private corporations.[29] U.S. investments in India complement these developments, as they seek access to top-level talent and India's large domestic market. Together, the investments being made on both sides demonstrate that U.S.–Indian collaboration in innovation is a "two-way street."

U.S. Private-Sector Investments in Indian R&D Facilities

U.S. high-technology corporations have initiated major investments in India to draw on India's innovation potential. Representatives from Google, General Electric, and IBM noted in their conference presentations that conducting world-class R&D in India is seen as a major opportunity to serve the Indian market while permitting more rapid development cycles that help them remain globally competitive.

Kenneth Herd of General Electric noted that the most compelling reason for General Electric to develop its R&D capability in India is the country's strong intellectual capital, embodied in its pool of talented engineers and scientists. By being on the ground in India, GE is able to take its place in India's vibrant innovation infrastructure, which includes more than 200 national laboratories and research institutes, 1,300 industrial R&D units, and over 300 universities with a strong student pipeline.

To capture these benefits, GE has invested over $80 million in the John F. Welch Technology Center in Bangalore. This facility, which houses state-of-the-art laboratories, employs over 2,500 engineers and scientists. To date the Welch Center has filed for over 370 patents, 44 of which have been issued. Mr. Herd noted that its R&D in India is on par with the world's best and that products developed in India, such as those related to diagnostic imaging, ultrasound sensing, and advanced plastics, help GE compete worldwide.

Ram Shriram, an Indian-born U.S. entrepreneur who is a founding board member of Google, similarly noted that Google believes that by leveraging India's innovation potential—in particular its scientific and engineering talent—it will help it to create new products such as Google Finance. Google has recently created two R&D centers in India, located in Bangalore and Hyderabad, which hire the top graduates from the Indian Institutes of Technology (IIT) and other

---

[29]As noted in the Preface, this 2006 National Academies' conference focused on India's emerging strengths in the automotive component manufacturing and pharmaceutical sectors rather than on India's more widely known accomplishments in the software and business service sectors.

leading schools in India. The reason to go local, Mr. Shriram said, was that not all talent resides in Silicon Valley, nor does everyone want to move there.

Far from considering India a labor-arbitrage, cost-saving destination, Mr. Shriram noted that Google views its operations in India to be on par with those at its headquarters in Mountain View, California. One indication of this parity is that the company's new global product, Google Finance, was developed by two researchers at its Bangalore laboratories.

Ponnai Gopalakrishnan of IBM noted that his firm is drawn to India not only for its skilled workforce, but also because India represents a very large market. India has unique market requirements, noted Dr. Gopalakrishnan, which call for designing systems and solutions for the high end as well as the mass market at the low end. Stating unequivocally that IBM's technological research and innovation in India matches the level and quality of its U.S. and other research laboratories, Dr. Gopalakrishnan noted that IBM's already large presence in India will grow even stronger with a planned $6 billion investment to develop a telecom research center in Delhi.

Finally, Eli Lilly's Robert Armstrong noted that India's embrace of intellectual property rights protections have not only created value for domestic industries such as Nicholas Piramal but also have enabled multinationals such as Lilly to consider undertaking high-end research and development activity in India. He noted that while the decision to outsource routine research functions to India was initially motivated by expectations of cost savings, Western pharmaceutical companies such as Lilly soon discovered that "embedded in that cost reduction is a lot of innovation in processes and focus on delivering products," which effectively reduces not only cost but the time to bring new products to market. Given the high cost and high uncertainties inherent in drug development, he said, developing high-end research and development capabilities in India, including alliances with Indian firms, are important vehicles for American pharmaceutical firms to improve the probability of technical and commercial success.

INDIA'S CHANGING LABORATORIES AND UNIVERSITIES

In addition to the rapid growth of corporate research laboratories, India's growing R&D capability is also reflected in the remarkable performance of its national laboratory system. According to R. A. Mashelkar, the director general of the Council of Scientific and Industrial Research (CSIR), India's national laboratories are rapidly evolving to link top-notch research with the needs of industrial competitiveness, social imperatives, and national benefits. This substantial improvements in the performance witnessed recently in India's national laboratory system is all the more significant given the size of this system. CSIR represents the world's largest chain of publicly funded industrial research and development organizations. It includes 38 national laboratories as well as numerous other research institutes located in all regions of India.

Dr. Mashelkar traced CSIR's positive transformation over a period of 10 years beginning in 1995 (when he was appointed as its director general) from an inward-looking organization to one significantly more forward in outlook and commercial in orientation.[30] Illustrating how CSIR is realizing its mission to serve India's people while advancing the commercial potential of research, he cited the development of a method of silver sulfadiazine microencapsulation on collagen-based biomaterial by researchers at India's Central Leather Institute. This technology, developed to advance India's leather industry, has been found to have a significant application in the treatment of burn victims. Another application that has promising commercial applications as well as addressing the needs of India's poor is a water filter developed by India's National Chemical Laboratory. This filter, the object of a 2005 U.S. patent, can screen out bacteria and even viruses from water at low cost. It is already being deployed across India to provide rural populations with safe potable water.

Dr. Mashelkar pointed out that CSIR's new emphasis on commercialization has not compromised CSIR's scientific credentials. As gauged by the Science Citation Index, the number of basic science papers published by CSIR researchers between 2001 and 2005 rose from 1,700 to 3,018 with one in six Indian papers published in internationally peer-reviewed journals coming from authors employed by CSIR's laboratories and institutes. Also, the number of U.S. patents granted to CSIR has increased from single digits through most of the 1990s to 145 in 2002 and close to 200 in 2005.

According to Dr. Mashelkar, numerous changes in the structure and administration of CSIR explain CSIR's dramatic improvement. To capture scale effects, the small projects of the past gave way to large, networked projects, with CSIR's many laboratories no longer behaving as independent entities. To foster a more commercial orientation, marketing teams were created in each laboratory, decision making was devolved, specialized businesses consultants were brought in, senior CSIR staff were allowed to serve on the boards of directors of private-sector firms, and awards were given for marketing and business development. CSIR also created financial incentives under a new strategy for intellectual property rights to motivate scientists, as well as established laboratory reserve funds that allowed laboratories to carry forward surpluses based on earnings that could then be used to fund additional research and development in future budget cycles. Underlining these changes, noted Dr. Mashelkar, was an effort to foster leadership in the laboratories by appointing individuals of exceptional merit who would stand tall in science, but who also have a realistic view of the continuity between knowledge and wealth creation.

---

[30]Dr. Mashelkar is internationally recognized for changing CSIR's bureaucratic culture to a more performance-driven, research and development organization over his 11 years as chairman. Among his numerous honors, Dr. Mashelkar was elected as a Foreign Associate to the National Academy of Sciences in 2005.

These accomplishments notwithstanding, Dr. Mashelkar argued that Indian attitudes about knowledge have still to evolve toward a more commercial perspective. Commercial spin-offs of the scale seen in Canada and the United States still do not occur in India. This may be due in part to a widespread cultural proclivity that separates knowledge—as represented by the Hindu goddess Saraswati—from wealth—personified by the goddess Lakshmi. Unlike in the United States, he said, "we have never understood the route from Saraswati to Lakshmi."

Challenges in Growing India's Knowledge Workforce

India also faces the challenge of growing the size and versatility of its knowledge workforce. Recently, a report by McKinsey warned that India will need an additional 1 million people to join the information technology and business process outsourcing workforce by 2010 in order to maintain its current market share.[31] This skills shortfall, the report warned, could threaten India's position as the leading offshore outsourcing location. As Dr. Mashelkar and others noted, expanding India's infrastructure for higher level technical education and preparing more students for the demands of the global workplace is a national priority.

The era of modern technical education in India began about 150 years ago, when the British established institutions of higher learning to train Indians to help its colonial administration. These were teaching rather than research universities.[32] Formal interest in technical education coupled with research emerged only after India won independence from Britain in 1947. India, in this postindependence period, made major investments in its scientific and technical infrastructure, including the founding of the Indian Institutes of Technology.

In his presentation, P. V. Indiresan, a former director of IIT-Chennai, expressed gratitude to the United States for its assistance in the 1950s in building India's university research infrastructure. He recognized, in particular, the contributions of Professor Norman Dahl of MIT who introduced a number of American institutional practices to the IIT system, including the semester system, continuous evaluation, and the credit system.[33] Technical cooperation offered by the United States also trained many among a new generation of scientists and engineers who on their return helped to build India's science and technology base.

The IITs today are internationally recognized as centers of excellence, not

---

[31]Domain-b, "Nasscom-McKinsey: India to Face Skilled Workers' Shortage by Next Decade," Dec. 12, 2005. Access at *http://www.domainb.com/organisation/nasscom/20051217_shortage.html.*

[32]Deepak Kumar, *Science and the Raj:* 1857–1905, New Delhi and New York: Oxford University Press, 1995.

[33]See Stuart W. Leslie and Robert Kargon, "Exporting MIT," *Osiris,* 21:110–130, 2006. Wishing to diversify India's educational portfolio by establishing IITs on several university models in addition to the IIT in Kanpur based on the MIT model, Prime Minister Nehru secured agreements for additional IITs in Bombay (in partnership with the USSR) in Madras (with the West Germans) and in New Delhi (with the British.).

least because they select students based on highly competitive merit-based admissions.[34] Only about 2 percent of applicants qualify for entrance—that is approximately 5,500 admissions out of 300,000 applicants.

Examining IIT's legacy, Professor Indiresan praised the institutes for producing in India world-class designers and analysts—many of whom are being sought after today by companies such as Google and General Electric in India and abroad. He noted, however, that they had done less well in innovation.[35] He suggested several reasons for this circumstance, including inadequate research budgets and cultural barriers (such as those noted by Dr. Mashelkar) that traditionally have segregated the intellectual castes (*Brahmins*) from the business castes (*Vaishyas.*) He also cited a complacent business mentality (shaped by long years under the License Raj) that, in the absence of meaningful competition, shunned innovation, though he affirmed that this attitude is now changing.

Liberalizing Education

Exploring IIT's future and the scope for India to expand the size of its globally competitive workforce, Professor Indiresan identified several constraints. First, he noted that heavy bureaucratic control by the Ministry of Science and Technology continues to stifle the potential of India's top research institutes and universities to expand and improve by growing their financial base. For example, he noted, the IITs are not allowed to accept donations directly from their alumni abroad, including a $1 billion gift that was offered at the height of the dot-com surge.

Professor Indiresan called for the government to deregulate the academic sector, much as it has liberalized the business sector, saying that this would encourage competition and innovation in India's higher education system. He noted that the IIT Act, which was inspired by the U.S. model, was the only act of the government of India that grants the level of autonomy that—ideally—all educational institutions in India should enjoy. Yet, even the IITs continue to depend on the government for their budgets since outside sources of funding are not normally allowed.

The Impact of the Reservations Policy

Second, he cited India's new reservations policy that seeks to address social inequalities by admitting students from traditionally underprivileged castes into the nation's elite institutions even though their marks fall below the minimum

---

[34]See Kanta Murali, "The IIT Story: Issues and Concerns," *Frontline*, 20(03), Feb. 1–14, 2003. Accessed at <*http://www.flonnet.com/fl2003/stories/20030214007506500.htm*>.

[35]For a discussion of the commercialization challenges facing IIT, see *Financial Times*, "India's Islands of Excellence Under Pressure," Feb. 21, 2003.

standards for admission.[36] Although this provides some poor but brilliant children an opportunity to rise, Professor Indiresan worried whether the IITs could maintain their world-class standards under this policy.

Promoting Excellence in Science and Research

Third, he noted that qualified senior professors—even at the IITs—are underpaid, commanding roughly the same salary as an intern in one of India's rising business firms. Poor pay leads to a short supply of qualified instructors and research investigators, with critical implications for the education of future generations of students. Professor Indiresan also called for an increase in university research budgets, noting that the small size of current allocations amounted to "making a suit from a six-inch length of cloth." Establishing science and technology parks of sufficient scale outside India's cities, where land is cheaper, he added, could also help support world-class research universities in India.

Expanding Opportunities for Quality Education

Finally, Professor Indiresan noted that India's elite technical institutions lack adequate competition at their level, which leads to a dulling effect on these institutes and universities in innovating their own structures and programs to meet the changing demands of globalization.

India's higher education system is two-tiered. Marquee institutions, such as the Indian Institutes of Technology and the Indian Institutes of Management, as well as selected private colleges, produce world-class graduates. Meanwhile, students at India's second tier of colleges and universities receive markedly inferior training. Legacies of the colonial model of education, many of these institutions still stress rote learning over the development of critical thinking, presentational, and problem-solving skills and teaming called for in today's employment market.[37]

To address India's emerging shortage of qualified graduates, Dr. Mashelkar stated that the country's entire system of science and engineering education is in the process of being overhauled.[38] Recalling that the IITs only accepted 2 per-

---

[36]This reservations policy is a topic of heated public debate in India at the time of the conference as the government of India planned to extend caste-based reservations to the country's premier universities and professional institutes, which previously had been exempt. See The Press Trust of India, "Government Preparing to Reply to Supreme Court on Quotas," June 16, 2006.

[37]Anand Giridharadas, "In India's Higher Education, Few Prizes for 2nd Place," *International Herald Tribune*, Nov. 16, 2006.

[38]These reforms, however laudable, thus far do not seem to address Indiresan's views with regard to the need for more fundamental reform, particularly with regard to greater privatization. The proposed reforms also do not appear to address the issue of easier market entry for foreign educational institutions.

cent of applicants, he noted that there remains an upper band among those not admitted to IIT who can show great promise if educated in a way that realizes their potential. This makes the case, he said, for upgrading many of the nation's better regional engineering colleges into National Institutes of Technology funded by the government of India. Other second-tier colleges and universities also are being targeted for improvement, Dr. Mashelkar noted, some with support from the World Bank. These and others, he added, could benefit from public–private partnerships that can link their curricula to the needs of India's increasingly buoyant private sector.

TOWARD A STRONGER U.S.–INDIA INNOVATION PARTNERSHIP

Greater understanding of India's changing innovation system—its challenges as well as opportunities—is needed if the United States and India are to realize the full scope of their new strategic relationship. An important theme of the conference reported in this volume is that the United States and India can both gain from a stronger innovation partnership and that there is significant scope for cooperation across many areas.

Strengthening Global Research Collaboration

The United States sees cooperation to improve India's knowledge infrastructure as enhancing its own innovation potential. As Thomas Weber of the National Science Foundation (NSF) noted, current research challenges are more complex than those of the past, requiring teamwork by researchers from various disciplines and from various parts of the world. International collaboration enables the United States, and India, to leverage resources that might not otherwise be available—data, experience, equipment, and infrastructure, among them—even as it furthers the U.S. goal of using its research grants to build a globally engaged scientific community.

"Global collaboration—among scientists, engineers, educators, industry, and governments—can speed the transformation of new knowledge into new products, processes, and services, and in their wake produce new jobs, create wealth, and improve the standard of living and the quality of life worldwide."

Dr. Arden Bement,
August 2005
Material Networks Symposium, Cancun, Mexico

Dr. Weber noted that greater collaboration between U.S. and Indian scientists is being stimulated by NSF study grants, by exchange of scholars between U.S. and Indian research institutions, and by links among research institutions. Currently, for example, students at MIT have the opportunity to spend a summer as a research intern at an IIT, complementing the more familiar phenomenon of Indian students studying and conducting research in the United States. International networking can also be facilitated through forums such as the Material Research Network that brings U.S. and Indian scientists together. The International Center for Materials Research at the University of California at Santa Barbara is, for example, strongly linked with leading Indian institutions including the Nehru Center, the Indian Institute of Science, and the Indian Institute of Technology in Mumbai.

Such collaborations can be extended and improved, Dr. Weber concluded, through improving (among other factors) broadband connections in India, agreeing on the rules of intellectual property rights and ensuring their enforcement, and reducing bureaucratic obstacles on both sides.

Growing Collaboration Across Innovation Systems

These ongoing real-world collaborations reflect an important way by which the United States and Indian scientists can collaborate to the mutual benefit of both countries. India's Science and Technology Minister Kapil Sibal noted in his keynote address that India needs to collaborate with the United States across a wider agenda, including an exchange of policy experience in the area of innovation and entrepreneurship. Such exchanges, he added, are in the interest of both countries.

Introducing Minister Sibal, Dr. John Marburger, of the White House Office of Science and Technology Policy noted that Americans would not be able to see the course of their future relationship with India by examining the trajectory of past interactions. The present differs too radically from anything known before, he cautioned, noting that new times require new approaches to collaboration. "We should be particularly eager to work with India, which is the world's largest democracy and is increasingly important to our own innovation economy, to magnify our mutual capacity to address our respective problems," he said.

In his remarks, Minister Sibal predicted that the 21st century will differ from the past in that a capacity to innovate in order to compete in global markets will increasingly determine the course of change and the wealth of nations. As nontangible intellectual assets become increasingly valuable in this 21st century paradigm, India's store of knowledge capital—embedded in its scientists and engineers—will be increasingly sought after. The challenge for India, he said, is to collaborate with the United States, while continuing to grow its own innovative potential.

India, he said, needs to collaborate with the United States to address global challenges such as energy sustainability as well as to address the needs of its poor.

To do this, the minister said, India is taking active steps to enhance its own innovation system. A product patent regime now in place is already spurring the pharmaceutical sector to switch from making copycat generic drugs to exploring new options such as making generic drugs for booming export markets and new drug research. Legislation along the lines of the Bayh–Dole Act has been introduced to Parliament, he said, with the goal of encouraging the commercialization of intellectual property found in India's educational institutions. In addition, the government plans to set up special economic zones and science and technology parks.[39] Also, a biotechnology development strategy is being implemented that will allow government funding for start-up companies. "We are going to public–private partnerships in a big way," he declared, "giving money to small- and medium-scale enterprises to make sure that they do new kinds of research for new products."

CHANGING INDIA'S INNOVATION SYSTEM

Economic liberalization has awakened Indians to their nation's potential, and particularly its capacity to innovate. This innovation potential is in no small part a result of major national investments made by India's postindependence leadership in setting up the Indian Institutes of Technology and the national laboratory system, among other core institutions. Drawing on India's substantial knowledge base, Indian industry has rapidly become internationally competitive in many sectors, while U.S. high-technology firms increasingly find it attractive to conduct advanced research and development in India.

Growth based on innovation has increased standards of living and reduced the number of Indians living in poverty. However, despite pockets of innovative activity in both formal and informal sectors, most innovation activity and productivity gains remain concentrated in a small segment of the Indian economy. As noted by several speakers at the conference, this dualism poses a serious challenge.

India's political and business leaders today recognize that for the country to continue its rapid development, new investments in the country's innovation infrastructure are necessary. While seeking to adopt global best practice with regard to policies and mechanisms that encourage pro-growth innovation, they are also aware that this changing innovation system must be inclusive if it is to be sustainable.[40]

The presentations at the National Academies' conference on India's Changing Innovation System underscore both the recognition of India's new window

---

[39]For an analysis of the potential and the challenges of these Special Economic Zones, see *The Economist*, "India's Special Economic Zones," Oct. 12, 2006.

[40]A multifaceted innovation agenda appears to be gaining currency in India. See Surinder Kapur, "Nurture New Technology and Innovation, Stay Competitive," *The Financial Express*, Nov. 16, 2006.

of opportunity as well as the significant challenges that must be overcome. As Montek Singh Ahluwalia noted, the agenda for changing India's innovation system includes continuing economic reforms through expanding the consensus on reform, and investing in the nation's hard infrastructure so that the benefits of reform touch all Indians. As R. A. Mashelkar and P. V. Indiresan affirmed, the agenda for changing India's innovation system includes a greater focus on commercializing the results of research conducted in India's universities and laboratories for commercial and social benefit, as well as expanding India's education base in a way that rewards merit while also being more socially inclusive. Growing India's manufacturing base also translates the benefits of a growing economy more broadly. As Surinder Kapur noted, this involves focusing India's business culture on quality production and practice, ready to adapt to new ways of doing things in order to be internationally competitive. Finally, as Minister Kapil Sibal observed, changing India's innovation system to meet India's development needs calls for enhanced cooperation with the United States, particularly given the growing interdependencies among the two large knowledge economies.

The United States can play a constructive role in facilitating the development of Indian capabilities by continuing to expand cooperative scientific exchange, demonstrating the value of a policy framework that facilitates the development and expansion of globally competitive R&D infrastructure, and, not least, by sharing best practices on innovation policies needed to unleash India's enormous pool of talent. This cooperation, in turn, can accelerate scientific advance and provide a positive environment for the expansion of the very real synergies that exist between two of the world's great democracies.

II

PROCEEDINGS

Welcome Remarks

Ralph Cicerone
National Academy of Sciences

Recalling the excitement that surrounded the successful visit to Washington of Indian Prime Minister Manmohan Singh in July 2005, the first month of his own tenure at the head of the National Academy of Sciences, Dr. Cicerone expressed his pleasure at the convening of the day's meeting, which was sure to afford a wonderful opportunity to explore the opportunities and challenges of greater cooperation in science and technology between India and the United States. India has a strong, rich, and productive creative tradition in science and mathematics of all kinds, so that while it is now a growing force in the global economy, it is far from new to science and technology.

Advances in information technology make it possible for the United States to benefit from other countries' innovative capabilities just as those others can benefit from America's. However, the benefits of globalization also pose challenges. Globalization pushes companies, individuals, and public institutions alike to adapt. One of National Academies' own recent reports, addressed to the U.S. Congress, stressed that with the pace of global competition increasing, the United States must adjust its policies and institutions if it is to compete in the future world economy.[1] And the United States is hardly alone in needing to adapt. Countries around the world are seeking to accelerate the transfer of scientific knowledge from universities, laboratories, and individuals into the marketplace. "In this process we must learn from each other," said Dr. Cicerone, calling that "the entire premise" of the conference.

---

[1]National Academy of Sciences/National Academy of Engineering/Institute of Medicine (NAS/NAE/IOM), *Rising Above the Gathering Storm: Energizing and Employing America for a Brighter Economic Future,* Washington D.C.: The National Academies Press, 2007.

From its "very amazing" list of attendees, dotted with the names of many distinguished Indians, he singled out for special welcome two of the country's ministers—Montek Singh Ahluwalia, Deputy Chairman of the Planning Commission of India, and Kapil Sibal, Minister of Science, Technology, and Ocean Development—as well as R.A. Mashelkar, who had been recently named a foreign associate of the National Academy of Sciences. He also recognized, representing the United States, Secretary of Energy Samuel Bodman; John Marburger, director of the White House Office of Science and Technology Policy; Under Secretary of State for Global Affairs Paula Dobriansky; Under Secretary of Commerce for Industry and Security David McCormick; and George Atkinson, science and technology adviser to the secretary of state.

Dr. Cicerone extended his appreciation for sponsoring the meeting to the National Academies' Board on Science, Technology, and Economic Policy (STEP Board) and to the numerous agencies that support it. Also singled out for thanks were Ram Shriram, a founding board member of Google, who had generously supported the day's event; the Confederation of Indian Industry, which had provided much help with its organization; and India's Ambassador to the United States, Ronan Sen, along with his staff. He then introduced Ambassador Sen.

Ronen Sen
Ambassador of India to the United States

Ambassador Sen noted that this conference comes as Indo–U.S. relations are rapidly transforming and a multifaceted strategic partnership is emerging, opening up new avenues for cooperation. He added that most of the joint initiatives between the United States and India, especially those of the previous two years, have been driven by science and technology. This list includes:

• The **Next Steps in Strategic Partnership:** This bilateral agreement, concluded in 2005, has extended the prospects for cooperation to civilian uses of nuclear, space, and dual-use technologies.

• The **U.S.–India High-Technology Cooperation Group:** This group is paving the way for commercial partnerships in information technology, biotechnology, nanotechnology, and defense-production technologies.

• A new, 10-year **Framework for the U.S.–India Defense Relationship**: Concluded in 2005 at the ministerial level, the framework has established a joint Defense Procurement and Production Group.

• A **Science and Technology Cooperation Agreement**: This landmark agreement was signed in 2005 by Indian Minister of Science and Technology Kapil Sibal and U.S. Secretary of State Condoleezza Rice.

• A new **U.S.–India Bi-National Science and Technology Endowment**

Fund: The purpose of this fund is to facilitate joint research projects with potential for industrial application.

- The **U.S.–India Knowledge Initiative on Agriculture:** An initial amount of $100 million, which has already been allocated, is bringing together research institutions and corporate entities in both countries for the purpose of raising agricultural productivity and increasing prospects for agro-industrial business in India.

- The **U.S.–India Energy Dialogue:** Launched in 2005 and cochaired by two of the day's speakers—Minister Ahluwalia and Secretary Bodman—not only envisages the vitally important **Agreement on Civil Nuclear Energy Cooperation** but covers such areas as oil and gas, clean-coal technologies, and renewable energy sources as well.

- The **U.S.–India Disaster Relief Initiative**: Adopted in the wake of successful joint tsunami relief efforts, this initiative is also technology driven.

- The **U.S.–India HIV/AIDS Partnership**: Established to tackle the disease on a global basis, this involves corporate entities in both countries in addition to the two governments.

- The **U.S.–India Working Group on Civil Space Cooperation:** This group is to renew and upgrade cooperation in space.

- The **Agreement on Civil Nuclear Energy Cooperation**: While benefiting both countries, this agreement is also expected to have a positive global impact, in part by helping develop more environmentally friendly and proliferation-resistant technologies.

In light of these many initiatives, as well as other recent initiatives listed, the Ambassador underscored that there could not have been a more opportune moment for a discussion of India's changing innovation system. Welcoming all to the symposium, he congratulated Dr. Cicerone and his colleagues at the National Academies on the event's timing.

Dr. Cicerone introduced the next speaker, Under Secretary of State for Global Affairs Paula Dobriansky, noting that she would be obliged to leave almost immediately after finishing her presentation.

Opening Remarks

India and the United States: A New Strategic Responsibility

Paula Dobriansky
Department of State

Dr. Dobriansky offered compliments to the National Academies' Board on Science, Technology, and Economic Policy for assembling such a distinguished group to address the critical issues under examination and a special welcome to Minister Ahluwalia, Minister Sibal, and Ambassador Sen. She conveyed Under Secretary of State for Political Affairs Nicholas Burns' regret for not being present due to a recent death in his family. On behalf of those assembled, she expressed her condolences and concern for his well-being.

Dr. Dobriansky noted that the United States and India find themselves at present in the beginning stages of what promises to be a very beneficial relationship for the peoples of both nations. Significantly, this partnership is not just between governments, nor could it be. Governments are not, after all, the creators of wealth, the makers of markets, or the source of human energy and ingenuity.

President Bush has remarked that India's greatest assets are its human resources and intellectual capital. More Indian students are studying in the United States than ever before, nearly 80,000 in 2006; for the third year in a row, India has sent a larger number of students to the United States than any other country, including China. The India–U.S. people-to-people network goes even deeper, however. Thousands of Americans live in Delhi, in Mumbai, and in Bangalore, while more than 2 million people of Indian origin, many of them now U.S. citizens, live and work in the United States.

The potential for U.S.–India relations, for years a topic of discussion, is finally being realized. President Bush's visit to India in March 2006 underscored the great progress the two nations have made in advancing a strategic partnership designed to meet the global challenges of the 21st century. This relationship, the

30

President has said, rests on the solid foundation of shared values, shared interests, and an increasingly shared view of how best to promote stability, security, and peace worldwide.

The United States appreciates that India is a rising global power. Within the first quarter of the new century, its economy is likely to take its place among the world's five largest. It will soon be the world's most populous nation. Its demographic structure bequeaths it a huge, skilled, and youthful workforce. It also continues to possess a very large and ever-more-sophisticated military force that is expected to remain very strongly committed to the principle of civilian control.

India and the United States are natural partners in confronting the central security challenges of the coming generation. The first, Dr. Dobriansky noted, is to gain and preserve access to sufficient supplies of food, potable water, and energy. The second is to counter terrorism and the proliferation of chemical, biological, and nuclear technology; international crime and narcotics; HIV/AIDS; and climate change. On these and many other issues, the two nations' interests converge.

A critical part of this blossoming relationship—discussed by President Bush and Prime Minister Singh during their March meeting—is the potential for cooperation in science and technology to improve people's lives. Many of the two nations' joint initiatives are based on this: the civilian nuclear initiative, the Agricultural Knowledge Initiative, the Bi-National Science and Technology Joint Commission, the clean energy initiatives, and the initiative to fight disease. Dr. Dobriansky proposed to highlight a few of these.

One of the most noted accomplishments of the President's visit is the announcement of a plan for moving ahead with the U.S.–India Agreement on Civil Nuclear Energy Cooperation. She called the plan to put the majority of India's nuclear program under International Atomic Energy Agency safeguards in perpetuity "truly an historic step forward." The U.S.–India agreement would remove an important source of discord that has affected the two nations' relationship for over 30 years, and at the same time enhance the international nuclear nonproliferation regime by bringing India further into its mainstream. It would also open up U.S.–India trade and investment in nuclear energy, thus helping India to meet its rapidly growing energy needs in a more environmentally friendly manner. The U.S.–India Energy Dialogue addresses other aspects of energy security by promoting the development of stable, affordable, and clean energy supplies.[2] To make possible full, peaceful civil nuclear energy cooperation and trade with India, Secretary Rice and President Bush have committed themselves to working with Congress to change U.S. laws and with the United States' friends and allies to establish an India-specific accommodation under Nuclear Suppliers Group Guidelines.

However, more than just the civilian nuclear initiative brings the two countries and peoples together. Prime Minister Singh has put economic reform at

---

[2]The cochairman of this commission, Secretary of Energy Samuel Bodman, discussed this initiative in his remarks at the conference.

the top of his agenda, and the U.S. agenda for developing deep economic and commercial ties with India had never been stronger. "Knitting together our two nations in a dense web of healthy economic interconnections," Dr. Dobriansky observed, is something from which both stood to gain.

Prime Minister Singh, in the role of finance minister in the early 1990s, had begun the process of opening India's economy to greater levels of foreign direct and portfolio investment, assuming a larger share of the world's trade by lowering tariff barriers, and creating a business environment that has sparked the development of a mobile telephone and an information technology and software services sector of world-class stature. It is a process to which Minister Ahluwalia remains a pivotal contributor.

This process is far from completed, however. To achieve sustained higher growth rates and broad rural development, India needs to further develop its airports, irrigation, and communication networks. It needs modern power grids, ports, and highways, as well as many other infrastructure improvements that could be vastly accelerated by greater investment, both public and private. U.S. businesses are even more likely to pursue opportunities in India as New Delhi presses ahead with privatization in areas ranging from insurance to power generation. Similarly, Indian labor market reforms and greater openness to foreign investment in the banking, retail, and services sectors would spark an enthusiastic response from American firms.

Advances had been made along several tracks of the U.S.–India Economic Dialogue during President Bush's March trip, highlighting numerous areas of bilateral cooperation:

Trade. The U.S.–India CEO Forum had made recommendations to President Bush and Prime Minister Singh to broaden bilateral economic relations substantially. The U.S.–India Trade Policy Forum is working to reduce barriers to trade and investment, with the goal of doubling bilateral trade within three years. Additionally, there is agreement on holding a high-level public–private investment summit in 2006 to work jointly toward completing the World Trade Organization's Doha Agenda before the end of that year. The U.S. State Department is very optimistic that these dialogues, with the continued and increasing interactions between the U.S. and Indian business communities, would contribute substantially to the prosperity of both nations.

Environment and Energy. The United States and India are working together to create focused government–industry environmental partnerships aimed at addressing shared environmental priorities, promoting activities with both local and global environmental benefits, and engaging the private sector in bilateral environmental cooperation activities. The U.S.–India Fund has underwritten more than 30 major wildlife conservation projects to assist India with conservation and management of its biodiversity, and new efforts are under way on U.S.–Indian collaboration to control illegal wildlife trafficking through the Coalition

Against Wildlife Trafficking. The U.S. Environmental Protection Agency (EPA) and Indian Ministry of Environment and Forests have signed a memorandum of understanding on cooperation on environmental issues. EPA and other agencies are working very closely with the government of India on issues ranging from air quality management to water resources and environmental governance. Both countries are committed to strengthening energy security and promoting the development of stable and efficient energy markets, and they are cooperating with four other nations in the region, through the Asia–Pacific Partnership on Clean Development and Climate, to promote the development of cost-effective, cleaner, and more-efficient technologies.

Governance. In 2005 the United States and India launched the Global Democracy Initiative to promote democracy and development. The two countries agreed to work closely in the region and globally to deepen democracy by offering their experience and expertise for capacity building, training, and exchanges to third countries requesting such assistance. India has also demonstrated the strength of its commitment to democracy by contributing $10 million to the U.N. Democracy Fund.

Agriculture. India derives 20 percent of its gross domestic product from agriculture, while more than 60 percent of its people made their living through agricultural enterprises. President Bush, recognizing agriculture's place in the lives and livelihoods of both Indians and Americans, visited the Agricultural University in Hyderabad and, with Prime Minister Singh, announced the revival of longstanding U.S.–India collaboration in agriculture. The Agricultural Knowledge Initiative is a three-year, $100 million commitment by the two nations to link their universities, technical institutions, and businesses to support projects in agricultural education, joint research, and capacity building, in the area of biotechnology among others. Designed to help India address its rural development and poverty issues through technology, research, and educational exchange, the initiative is a high priority for Prime Minister Singh and a symbol of strong, shared commitment to rural development.

Disease Control and Prevention. The two governments are working very closely to confront the major challenges of HIV/AIDS and avian influenza. The U.S. Centers for Disease Control and Prevention are working with India through the CDC's Global AIDS Program. In addition, the National Institutes of Health is supporting Indo–U.S. collaboration in HIV/AIDS research in such areas as the development of vaccines. Meanwhile, to boost private-sector involvement, an Indo–U.S. Corporate Fund for HIV/AIDS has been established. In the field of avian influenza, India is demonstrating world leadership: It has been among the first nations to join the International Partnership on Avian and Pandemic Influenza and has agreed to host the Partnership's global conference in 2007.

Science and Technology. Current collaborations between the United States and India in science, technology, engineering, and related research and development are genuine partnerships, not merely the assistance programs of the past.

Both Indians and Americans have long been recognized as being leading innovators in information technology, biomedical research, biotechnology, agriculture, and many other high-tech fields. "India has the scientific and technological base to join the United States as equal partners in pushing forward the frontiers of research," said Dr. Dobriansky, something that would result in "a very positive impact on the lives of all of our citizens." Indeed, a key to mutual economic growth and prosperity is to increase linkages among U.S. and Indian knowledge bases: the two nations' scientists, engineers, researchers, academics, and private sectors.

A new Science and Technology Framework Agreement signed in fall 2005 by Secretary Rice and Minister Sibal establishes, for the first time, intellectual property rights protocols and other provisions truly necessary for conducting active collaborative research. The agreement also builds the framework within which Indian and U.S. scientists in government, the private sector, and academia can collaborate very actively in such areas as basic and engineering sciences, space, energy, health, and information technology.

For many, these new opportunities for increased scientific collaboration come as no surprise. Scientific and economic links between India and the United States have been strong since the very early 1960s—first in agriculture, then spreading into a broad range of areas involving most U.S. governmental agencies. The benefits are currently visible in many parts of the United States, where many experienced Indian scientists and engineers are working as a result of such active collaboration.

Under this S&T Framework Agreement, the United States and India would cofund a $30 million Bi-national Science and Technology Endowment Fund that is designed to generate collaborative partnerships in science and technology, as well as to promote industrial research and development. In addition, the United States and India are exploring the potential for cooperation in Earth observation, satellite navigation and its application, space science, natural hazards research, disaster management and support, and education and training in space. U.S. instruments are to be provided for India's upcoming first lunar mission, the Chandrayan 1; at a time when the United States had not gone to the moon for many years, this represents an opportunity for the two nations to collaborate on efforts to understand Earth's closest neighbor.

Despite the number of initiatives listed, said Dr. Dobriansky, she felt as if she had merely scratched the surface of Indo–U.S. collaboration. Calling the countries "natural allies [who were] finally realizing the full potential of close corroboration," she declared that with the help of the innovators, scientists, and entrepreneurs in the audience, the benefits of U.S.–Indian friendship would be felt by all among both peoples, from farmers to physicists. She concluded by expressing her support for the day's discussions, her eagerness to be apprised of their outcome, and her thanks for having been given the opportunity to address the symposium.

India and the United States:
An Emerging Global Partnership

Moderator:
David McCormick
Department of Commerce

Dr. McCormick, describing the panel's topic as ambitious and exciting, said he would be very brief in introducing its distinguished presenters. He then invited Montek Singh Ahluwalia, deputy chairman of the Planning Commission for India, to speak on the topic of "India's Reforms: Current Challenges and Opportunities." The deputy chairman, he said, has a long history of public service in a variety of leadership positions in India, and, more recently, at the International Monetary Fund. A prolific scholar, he is not only a student but also an architect of many of India's key economic reforms of the past 25 years. Offering thanks for Mr. Ahluwalia's participation, Dr. McCormick turned the microphone over to him.

INDIA'S REFORMS:
CURRENT CHALLENGES AND OPPORTUNITIES
Montek Singh Ahluwalia
Planning Commission of India

Mr. Ahluwalia began by proclaiming his delight that an institution accorded such prestige in India as the National Academies had seen fit to cosponsor a forum for discussing the scope for economic cooperation between India and the United States—a matter whose importance is growing rapidly. As the summary of ongoing cooperative activities provided by Dr. Dobriansky had been comprehensive, he would limit himself to expressing his contentment that collaborations between Americans and Indians are deepening and to observe that this trend signals a welcome and exciting change in the relationship between the two nations.

Basic Premises of Indian Economic Reforms

Mr. Ahluwalia noted that his talk would focus on the economic reforms that have been critical to altering perceptions in the United States and elsewhere of the state of India's economy and of its relevance to the world economy. In their content, the Indian reforms have not differed greatly from those implemented in many developing countries. These reforms reflect India's acceptance of four basic premises about a strategy for growth:

1. **Private enterprise is a critical driver of growth.** The recognition that India's private sector, which Mr. Ahluwalia rated as very strong, deserves support and encouragement has been critical to the reform effort. In his judgment, the country is quite capable of taking on competitors, provided the playing field is level.

2. **Competition spurs efficiency.** Although India does not need to create a private sector, its reforms are in great measure designed to increase competition within the private sector already in existence.

3. **An open, integrated economy is preferable to a closed, insulated economy.** There have been a number of initiatives aimed at opening India up to both trade and foreign direct investment.

4. **India's private sector should be encouraged to seek opportunities abroad.** Indian companies have begun looking at both new investments and acquisitions of companies offshore, a major change in the country's economic environment that is creating a far more symmetric kind of globalization.[3]

In all four areas, India's reforms have taken a gradualist approach, influenced by a pair of factors. The first is the strategic perception that it is better to exercise caution in moving forward than simply to undertake shock therapy. The second is a deliberate decision to move forward at a pace that would build consensus for change, thereby avoiding excessive controversy over any one issue. Mr. Ahluwalia reminded the audience that India is not only the world's largest democracy but also a "very pluralist" one, that for the previous decade it had been run by coalition governments, and that the governments of its states were in the hands of a variety of political parties. As proof that a consensus in favor of change had been achieved, he cited the contrast between the national debate, which might often strike students of Indian politics as being at least somewhat contentious,

---

[3]See, for example, *The Economist*, "India's Acquisition Spree," Oct. 12, 2006.

with the fact that the many state governments are moving in the same direction of reform. The conscious choice of gradualism has aroused a fair measure of irritation and impatience among many of India's friends abroad—feelings that, he acknowledged, are "not unknown even to those of us in India who have to deal with this subject"—but this was to be regarded as one price of building a broader consensus, which has proved effective.

What enables him to say that India's economic reforms were essentially on solid footing? Above all, the results they are producing. The economy's annual growth rate, consistently above 7 percent over the previous half-decade, reached 8.4 percent in 2005. The country, targeting continued acceleration, hopes to push its growth rate to 9.5 percent per annum by the end of the following five years while achieving an annual average of 8.5 percent over the entire period. That would put India on a growth trajectory comparable to those that South Korea and China enjoyed in the previous two to three decades. There was agreement both within India and among international observers that growth on this order is "feasible" as long as India followed what the deputy chairman called "the right policies." The Planning Commission had, shortly before, sent to state governments a discussion document laying out very detailed prospects for the five years to come. "We are trying to emphasize here," he stated, "that these transitions are not automatic."

Four Major Challenges Facing India

He then listed challenges that India is facing in four critical sectors—agriculture, social services, infrastructure, and energy—and declared that each offered a very substantial scope for advancement through cooperation between India and the United States.

Agriculture

A very large percentage of India's population continues to derive the majority of its income from agriculture, even though agriculture's component of India's GDP, at 20 percent, has fallen off significantly. Indian planners are aiming for a "second Green Revolution" to transform the nation's economy further. While the goal of the initial Green Revolution was to produce enough food to meet the country's needs, the Second Green Revolution is to be focused on achieving broad-based income growth in rural areas. Its foundation is to be a high degree of diversification in agriculture, which implies the growth of agro-based processing activity and greater efforts in modern marketing than have been made to date. Technology is expected to play a crucial role in this wide-ranging change.

The minister then pointed to evidence that change has already begun. Remote-sensing satellites are playing a role in water management as India attempts to cope with moisture stress in the two-thirds of the country that lack assured irrigation. Information from the satellites is being used in methods of land development designed to preserve water; in the planning of local-level irrigation systems, for instance, it is guiding the placement of small dams that retain water. Stressing the importance of the knowledge initiative in agriculture referred to by Dr. Dobriansky, he said that India hoped to rebuild technical and scientific linkages between U.S. universities and its own, calling such connections crucial for his country's future. Among areas for inclusion, he suggested, were moisture stress and technologies related to food processing.

Social Services

A major focus of India's Eleventh Plan is to build an infrastructure for meeting basic education and health needs, and, beyond that, "hugely" strengthening the country's capacity to provide the skills required for integrating with the global economy. The country's system of higher education, which historically had been very well developed, is capable of producing a wide range of skills, as reflected in the comparative advantage India enjoys in such fields as information technology (IT) and IT-enabled services or research and pharmaceutical biotechnology. Nonetheless, skill constraints have become evident with the expansion of the country's economy. While India's declining dependency ratio puts it in an advantageous position relative to the industrialized nations and even to China, this "demographic dividend" would only be capitalized if India can convert it into very highly skilled manpower. Major changes in the educational system are needed both to expand education and, even more important, to ensure quality.

Infrastructure

It is the consensus view among policy makers in India that high-quality infrastructure is a critical requirement if India is to achieve its desired annual growth rate of 8.5 percent. A number of initiatives are under way in the areas of seaports, roads, railways, airports, and electric power. All are designed to spur a major increase in investment and to bring in new technology, with public–private partnership being one of the tools. Results have been favorable to date, he added.

Energy

The Planning Commission believes that, to sustain an economic growth rate of 8.5 percent, India would need growth in total energy supplies on the order of 6.5 percent per year, a figure that assumes great improvements in energy efficiency. Underlining the potential this might offer for collaboration between India and the United States, Mr. Ahluwalia noted that he and U.S. Secretary of Energy Samuel Bodman, who was seated with him on the dais, cochair the Indo–U.S. Energy Dialogue. This dialogue envisages significant cooperation in many areas of energy efficiency and in such clean-coal technologies as *in situ* coal gasification and coal-bed methane exploitation.

The Indo–U.S. nuclear agreement, recently signed by President Bush and Prime Minister Singh, holds out the possibility of a major transformation of bilateral cooperation in exploiting nuclear energy's potential. Nuclear energy could greatly aid India in reducing its dependence on coal.

India also expected to see increased cooperation with the United States in the extraction of ethanol from agricultural waste, as well as in many other energy areas that it was interested in pursuing.

The "bottom line" of the nation's economic reforms is the fact that competition has put Indian industry under tremendous pressure to embrace innovation and change. This was evident in the way individual Indian companies are defining their corporate strategies and in the increased investments they are making in technology acquisition and upgrades. It is visible as well in the government's putting in place an intellectual property regime consistent with the new environment.

Success Paves the Path of Reform

India's transition has been associated with a perception of success: The nation's industry has benefited hugely from the reforms, and many companies have positioned themselves in a way that has increased confidence, bolstering the government's conviction that it should move forward. The deputy chairman wished to reassure the audience that the Planning Commission judges the reforms to have worked well and that the government would continue down the path of reform. He hoped that debate within India on how to realize the desired rate of growth would greatly intensify in the months ahead as a result of the release to the state governments of the previously mentioned discussion document.

Having thanked Mr. Ahluwalia, Dr. McCormick introduced Secretary Bodman as ideally suited to speak on U.S.–India Science and Technology Cooperation as a scientist, scholar, former CEO of a Fortune 500 company, and current senior government official.

OPPORTUNITIES AND CHALLENGES
IN U.S.–INDIAN SCIENCE AND TECHNOLOGY COOPERATION

Samuel Bodman
Department of Energy

Underlining the importance of the day's meeting, Secretary Bodman expressed gratitude for the remarks of Deputy Chairman Ahluwalia, whose leadership in India's economic reform effort he called "a major reason that we now see India both as a potential partner and a competitor." India is a friendly competitor, to be sure, but an effective competitor in the global marketplace nonetheless.

Secretary Bodman observed that, with energy issues currently at the forefront in the United States, the press of business would require his return to the Energy Department immediately following his talk. Apologizing in advance for his early departure, he stressed that the importance of the U.S.–India relationship was such that he had wanted to be sure of a chance to hear Mr. Ahluwalia speak and to say a few words himself about the benefits of U.S.–Indian cooperation in science and technology, which he called "a subject near and dear to my heart."

A New Era of Cooperation and Trust

Cooperation between the United States and India is crucial to the global marketplace, as it has been to the spread of the democratic model of governance throughout the world. Both he and President Bush were very pleased that a new chapter has been opened in the relationship between the two countries, one based not only on mutual needs but on an increasing level of trust. As the President said in March 2006: "Our two great democracies are now united by opportunities that can lift our people. The United States and India, separated by half the globe, are closer than ever before, and the partnership between our free nations has the power to transform the world." Secretary Bodman agreed with those sentiments and believed that great power comes from mutual interest in science and in technology.

The Secretary said that he began his adult life teaching at MIT, and many of the very best students there were from India, a situation he was sure still stood. In 2005, President Bush and Prime Minister Singh declared their resolve to transform the relationship between their countries in ways that would support and accelerate economic growth through greater trade, investment, and collaboration on science and technology issues. This cooperation will do much to enhance energy security for both countries because it can promote the development of stable and efficient energy markets and enhance the research and development of alternative energy sources, work that was already under way.

Collaboration Under Way in Energy

Their joint statement also referred to the International Thermonuclear Experimental Reactor (ITER), a partnership dedicated to developing a facility for demonstrating the technological feasibility of fusion energy. India has joined six other nations in initialing the ITER agreement in May 2006; together, the seven parties to the agreement represent more than half the world's population. Through ITER, India will be playing a very important role in harnessing fusion as an inexhaustible source of pollution-free energy for the world. While allowing that success in this enterprise will not come during his own tenure at the Energy Department, Secretary Bodman expressed the hope that it would make life easier for his successors. The United States also welcomes India's collaboration on the development of the proposed International Linear Collider (ILC), which is expected to make possible new discoveries in particle physics. The ILC is to be designed, funded, managed, and operated as a fully international scientific project. "I hope we can attract the interest of India to participate," he said.

The Science and Technology Cooperation Agreement that the United States and India signed in October 2005 has established a framework for the exchange of ideas, information skills, and technologies. Under its terms, the countries will be able to advance scientific progress in clean-energy research and development, in the sharing of training facilities, and in the exchange of materials and equipment. An example of the type of cooperation that the United States hopes to encourage is India's decision to join the FutureGen international partnership, an effort to create a zero-emission, coal-fired power plant that would convert all the energy in the coal used to fuel it into a stream of clean hydrogen while sequestering the resulting carbon dioxide beneath the ground. Secretary Bodman noted that the United States appreciates India's agreement to participate in both the government steering committee guiding the project and the industry alliance handling its actual construction, as well as its pledge of $10 million in financing. This major investment in future technology, he hoped, would benefit the entire world.

The two countries are also working together to bring India into the Integrated Ocean Drilling Program, as well as cooperating on the study of methane hydrates, or clathrates. India's work on the latter would involve U.S. technology, which the Department of Energy was very happy to provide. Many researchers from both nations are to take part in this and follow-on efforts, allowing acceleration of the commercial utilization of hydrates in the United States and around the world.

The U.S.–India Energy Dialogue, mentioned earlier by Mr. Ahluwalia, has led to the creation of five working groups:

- **The Civil-Nuclear Working Group** has already held one technical workshop to advance the dialogue between the two countries; a second workshop is to take place in the United States later in 2006.
- **The Power and Energy Efficiency Working Group** has a varied portfolio. Under its auspices, the U.S. Agency for International Development and General Electric have formed a public–private partnership whose goal is to provide up to four rural communities in India with access to clean and affordable energy over the following two years. It endeavors to establish avenues for technology cooperation on for industrial- and building-energy efficiencies. Just the previous month, a conference in Delhi—aimed at spurring business partnerships that would result in the application of new, energy-efficient technologies—attracted significant participation from representatives of both countries' governments and business communities. Still under consideration is a strategic partnership between India's National Thermal Power Corporation and the U.S. Department of Energy's National Energy Technology Laboratory that would advance the development of clean, efficient power generation.

• **The Coal Working Group** has created a high-level work plan identifying priority projects for the next couple of years, which includes the pursuit of investment opportunities and information exchanges in the areas of coal mining and processing, coal mining safety, and *in situ* coal gasification.

• **The Oil and Gas Working Group** held a one-day, government-to-government workshop in New Delhi the preceding month to pursue development of a regulatory framework for natural gas; ways of involving the U.S. and Indian business communities were being explored as well. Several bilateral Memoranda of Understanding have been arranged through the Working Group; they cover information exchange, operational safety, inspection issues, and investigation of accidents related to both drilling and production activities of offshore oil and gas operations.

• **The New Technology and Renewable Energy Working Group** is sponsoring meetings between the two governments to discuss potential areas of collaboration, including solar power generation, low-wind-speed technology, and other renewable energy resources.

Secretary Bodman reiterated his belief that the partnership in science and science-related matters that is emerging between the United States and India would be of great benefit to both countries. President Bush, in his 2006 State of the Union Address, had laid out what the Secretary described as an ambitious but achievable program to expand research and development in alternative resources of energy. Known collectively as the President's Advanced Energy Initiative, these efforts are geared to bring to the market energy produced using cellulosic ethanol, hydrogen, solar, and wind-based technologies. All require collaborations among the very best scientists and engineers in the world. While many of these individuals are to be found in the United States, the Secretary stated that they are surely to be found in India as well. He expressed his own hope that the two countries would achieve the same level of cooperation on these projects as on ITER and on FutureGen, which would make it possible to bring them to reality that much sooner.

In closing, the secretary expressed gratitude for the invitation to speak at what he called an "important meeting." He expressed his appreciation for the spirit of cooperation and noted further that the meeting was taking place, at the National Academies—"the heart of the American science community." The conference was thus symbolic of a new and, he hoped, very prolific chapter in the history of U.S.–Indian relations.

Dr. McCormick, thanking the secretary, opined that the web of collaborations and initiatives that had already emerged in the comments of the day's first few speakers was truly remarkable. The avenues of the two countries' cooperation were both many and exciting.

He then introduced Ram Shriram, who has been associated as both an operating executive and an investor with some of Silicon Valley's greatest success stories of the previous two decades. From his experience as a founding board member of Google and with other leading innovators—Netscape and Amazon among them—he was uniquely qualified to describe new models for U.S.–India innovation and for collaboration.

NEW SYNERGIES IN U.S.–INDIAN COOPERATION

Ram Shriram
Google

Noting that he was an outsider to Washington but a Silicon Valley insider over the previous 25 years, Mr. Shriram proposed to share some personal experiences, most specifically those with Google—in particular, how Google uses innovation as a competitive advantage to build its business. He said that he would then evoke one or two specific examples of innovation-based new products that have come out of the company's laboratories in India.

Far from involving magic or mystique, Google's success is based on a "rather simple" formula that begins with hiring "really, really smart people." From the earliest days of the company, its two founders, Larry Page and Sergey Brin, decided not to compromise on hiring high-quality employees. The first engineers Google hired were Ph.D.s, some from Stanford, the rest from other U.S. institutions. (Although neither Page nor Brin, Ph.D. candidates when they met at Stanford, had completed the degree, Mr. Shriram assured the audience that their former professors rated them among the brightest in their cohort.)

Google's Blueprint for Innovation

Google's features and practices aimed at maximizing innovation include:

- A **flat management structure,** making for easy communication up and down, thus reducing confusion within in the organization.
- **Encouraging constructive chaos** while keeping teams small and nimble so that their projects were "very measurable and very doable,"
- **Avoiding silos** in keeping with the company's open, communicative environment.
- An **"Ideas Mailing List"** on an intranet running inside the company's firewall that allowed employees to mail their ideas in.
- **Offering engineers 20 percent of their time** to work on anything they wanted as long as this did not compromise project timetables. Mr. Shriram called the practice "extremely effective," asserting that most creative people appreciate the flexibility of not being "chained" to a single project, whether they are employed by an established corporation or a start-up.

- An **iterative design** process, anchored in the belief that the best ideas don't make their appearance all at once, but emerge over time. He described most company decisions as "highly data driven," based on feedback from the more than 150 million people per day who visit Google's Web site, thus providing it with a "huge laboratory" in which it can "refine and define" its products.
- **Server based deployment** in huge data centers, six in the United States and many more overseas. "We have one instantiation of the code that we run, which is what you access as Google products at home on your PCs or on your handheld mobile phones or other devices," he explained.
- **"Test, Don't Guess,"** the philosophy behind Google's constant iteration and improvement, as well as the reason Google products may carry the label "beta" for a year or more. "We're proud that we are in beta," Mr. Shriram said, "because we really don't want to release products as full-fledged till we feel they're ready for prime time."

Summarizing what he called Google's "excellent hiring process," Mr. Shriram said that, because of the pace of technological change in its field, the company does not believe in recruiting experts. Rather than looking for a specialist in, say, storage or user-interface design, Google prefers to develop someone it had hired on the basis of raw intelligence and willingness to learn. The company is adding about 15 employees a day, most of them highly skilled: Of a total workforce of around 7,000, 2,500 were engineers, and 500 of those hold Ph.Ds. "It's important for us to be able to scale the level at which we're hiring without making any serious mistakes," he said, adding that this was accomplished through an Applicant Tracking System.

Getting a job at Google began with solving a mathematical puzzle. Posted along U.S. Highway 101 in Silicon Valley is an algebraic equation whose solution took prospective applicants to a Web page where their application would be accepted. With resumes coming in at a rate exceeding 20,000 per day, the company felt it could not handle the volume otherwise. "We are doing our best to filter applications out so that we can hire the people we want while not wasting the time of those who may not be the best fit for our corporation," Mr. Shriram said.

Company Organization Accents Transparency

In its internal processes, Google does everything that many major corporations do—delegating document review, for example—but in what he described as "a very lightweight sort of online, intranet way." All projects were posted on the company intranet for everyone to see, so that even small projects were called to colleagues' attention. "We share everything, we talk about everything, and we use our own products internally as a way to communicate as well," he added.

Google's engineering leadership has a narrow structure: three or four people at the top, with a large number of project leads across the company. Employees are often moved from project to project, which allows them to become well versed in such areas as mobile design, user-interface design, or internationalization, all of which the company considers important. "When you cross-pollinate," Mr. Shriram stated, "you develop really well-rounded people with a lot of skill sets and a lot of talent."

Thinking Globally from the Start

Seeing its footprint as global, the company takes a global view of each of its projects from day one. Google Maps, for example, was not limited to maps of the United States; the company viewed it as imperative to include maps of Europe, Asia, and elsewhere as quickly as possible. Indeed, of Google's current traffic, only 30 percent originates in the United States, and the company expects the 70 percent coming from overseas to grow, with India and China expected to become the largest points of origin within 10 to 15 years.

Mr. Shriram said he has been a major proponent of Google's expansion in India, and that the company's two founders had been thrilled with their experience of seeing the country up close during a 2005 visit. Google's efforts in India began with an R&D center in Bangalore whose staff has since grown to about 60 from 5 or 6 at its founding three and a half years before. The company currently has two Indian locations, the other being an operations center in Hyderabad employing around 350. Finding people for the Bangalore center who could "be easily assimilated into the Google development environment" initially proved a challenge, because the company had not gone about it the right way. Realizing that it needs to recruit at the universities, however, Google has since succeeded in hiring top graduates from Indian Institutes of Technology and others among "the best available researchers" in the country. The company had sponsored a pair of programming contests, the Google India Code Jams, which had drawn 25,000 participants. "As much as people think that there's a lot of talent available in India," he cautioned, "it is a very competitive environment, so you have to be out there seeking that talent directly at the source."

There is a very strong belief inside Google that all its R&D centers should be equal. Far from considering India a labor-arbitrage, cost-saving destination, Google views its Indian operations to be on a par with those of its Mountain View headquarters or the R&D centers it had opened in New York City, Zurich, Tokyo, or, most recently, at Beijing's Tsinghua University. Google recognizes that not all talent resides in Silicon Valley, nor does everyone want to move there. "We need to have a global view, not a narrow, Silicon Valley-centric view," said Mr. Shriram, adding that Google wished to avoid "the hubris that often develops in a successful company."

From Bangalore, a Promising New Product

A sign of its success in hiring "Google-caliber" engineers in India is the development of an interesting new product, Google Finance, by two researchers at its Bangalore labs. During their "20 percent time," they came up with the core innovative concept of flagging the occurrence of news events along moving charts that track stock prices. From an engineer in Google's New York office, a young Romanian "who was well versed in macromedia flash but didn't know the difference between the NASDAQ market and a grocery market," came the Ajax code that allowed the company to launch this sort of moving chart. At present, Google has 10 engineers at three labs across five different time zones working on the product, an example of cooperation at the global level within the company.

Google Finance itself, which was still in beta, or testing, phase, is in fact quite complex, in that all its graphics have to be sent down to a given computer fairly quickly—whether the device is on a broadband or a narrowband connection—so that the user can see the movement of the chart when pulling a news item into or out of it. Mr. Shriram classified it as a "mainstream product," saying it has been very well received by U.S. users, who are particularly interested in its application to stocks traded on the NASDAQ or New York Stock Exchange. Urging those members of the audience who had not played with Google Finance to try it at home, he called it "as good as having a Bloomberg terminal on every desk."

Deepening U.S.–India Ties: A Private-Sector View

Mr. Shriram then said that he wished to share his own perspective—that of one who is active in the private sector—on ways in which the U.S.–India partnership might be deepened. A major priority is to attract a larger number of U.S.-trained Ph.D.s of Indian origin back to teach at the Indian Institutes of Technology (IITs) and Indian Institute of Science. Building a culture that rewards innovation across the spectrum of Indian educational institutions, not only by improving levels of pay and recognition for professors but also by providing them a platform for growth, is needed to create this dynamic.

Access and affordability for prospective students are also priorities. Increasing the number of admissions at the IITs is important because, in comparison to the 10 percent or so of applicants to the U.S. Ivy League universities who successfully graduate with degrees, only around 1 percent of the IIT applicants actually make it all the way through. While Mr. Shriram was willing to wager that some percentage of those rejected by the IITs have been subsequently accepted by Ivy League schools, he posited that there were many applicants who were nearly selected and who could handle the coursework and benefit from alternatives in high-quality education. Offering additional direct incentives for investment in education and research in India would therefore be a very proactive, helpful step.

Innovation drives the growth and success of companies as it does of nations, and if there is going to be innovation in the product sector of software companies, more computer science graduates will need to come out of these IITs than is the case currently. In view of the prevailing shortage of skilled people and of the number of both U.S. and Indian companies hiring science talent, the supply side demands immediate, collective attention. "We know the talent exists," he said; "they just don't have the ability to go to school."

Furthering Exchange of Skilled Personnel

Highly desirable as well is additional "joint level 'measurable' research" with support from U.S. corporations, especially focused around the educational institutions. Mr. Shriram said that he is urging Google to partner with more institutions of higher learning in India. Whether this strategy will ultimately reward Google by swelling its ranks with skilled employees is open to debate, but it is certain that this initiative will nurture talent in India by providing promising researchers the opportunity to take their work further. At the same time, more U.S. students should be invited to spend a year in the Indian environment, either with U.S. corporations that are based in India or with Indian firms. Although this is already becoming a trend, it merits encouragement because it is an excellent way to augment communication between the two countries at the grassroots level.

To conclude, Mr. Shriram emphasized that innovation without execution represents wasted effort, evoking the example of Xerox PARC in the 1970s and 1980s. Some of the brightest Ph.D.s in the United States worked at PARC, and a great deal of innovative thinking took place there. Yet, while the innovation from PARC benefited the country as a whole through the products of other firms such as Apple Computer and Microsoft, Xerox itself never benefited sufficiently from its investments in PARC. It is therefore important to try to ensure the existence of a virtuous cycle from innovation to successful product to profit, with the profit then plowed back into innovation and, perhaps, education and research. It is this successful cycle that fosters more innovation.

Dr. McCormick thanked Mr. Shriram for providing a fine overview of Google's activities, in particular those related to India. The company's global reach had become apparent to him several weeks earlier during a visit to China. A senior executive of a Chinese company began asking, in the course of a lunch, some very detailed questions about the U.S. Constitution, how it had been developed, the history of the Bill of Rights and other amendments, among various other details. Taken aback by these questions, Dr. McCormick asked: "How did you become so conversant in this topic?" The answer: "I've studied this in great detail on Google." The extent of Google's reach was significant, therefore, and its implications much broader than often imagined.

DISCUSSION

Dr. McCormick, remarking that many interesting topics had been raised by the panel's three speakers, then proposed to devote the 15 minutes that remained of the session to questions from the audience.

Developing India's Information Infrastructure

Harvey Newman, a professor of physics at Caltech, said that he wished to pose a question to both Mr. Ahluwalia and Mr. Shriram, but he felt that it first needed to be put into context.

Dr. Newman said that as chair of the Standing Committee on Regional Connectivity, which deals with global collaborations, he has worked on the issues of network development, grid development, and digital divide in many different regions of the world. Since the focus on India has been very strong of late, he said that he noted with specific interest Mr. Shriram's statement that there soon would be a great deal of network traffic and communications involving India and China. This had also called to his mind a number of recent speeches by India's president, A. P. J. Abdul Kalam, articulating a vision of a grid for a billion people in India.

Despite these optimistic predictions, however, India remains about 10 years behind the more advanced regions of the world, and its transition to modern telecommunications has yet to begin. Dr. Newman wished to hear Mr. Shriram's perspective on this issue, as well as Mr. Ahluwalia's view on the next steps to be taken in developing this infrastructure, which would provide students and researchers throughout India with access to information.

Responding first, Mr. Shriram said that many in India's private sector are already working on broadband infrastructure. A major example is Reliance Corporation, which is putting in a backbone fiber network. However, he warned against looking at India through a U.S. prism, saying, "They don't have the exact infrastructure that we have." The last mile in both India and China is likely to largely go wireless, permitting the bypass of the entire generation of copper-wire technology, with mobile phones likely to provide the initial vehicle. In those two countries the phone is currently incorporated in the PC for many communications, both text and voice, and multimedia messaging services might later be added. From 4 million to 5 million new mobile connections were being made each month in India, and even if not all are on the high-speed network represented by the GPRS platform, in time they would be.

Mr. Ahluwalia, endorsing Mr. Shriram's viewpoint, stressed India's awareness that it is "absolutely crucial" to develop its telecom infrastructure and make available ICT connectivity. Indications from the previous five years have all been very positive, and although there is indeed less connectivity than the country would like, current talk is of possible increases in multiples. Besides the private-sector players, active government-sector companies are involved, and they are in fact rolling out quite a substantial network. The deputy chairman shared Dr. Newman's perception that the process needed to be accelerated if India is really to catch up and said he is unaware of "any major policy hassle" holding up connectivity, although "individual bits and details" might have to be overcome.

Expanding Capacity in Higher Education

Anant Narayanan, introducing himself as an alumnus of an IIT and a lawyer, said that in the former role he had been involved to some degree in discussions relating to the expansion of higher-education capacity as alluded to by Mr. Shriram. Speaking on behalf of his fellow IIT alumni and of those harboring good wishes for India and its future, he said all realized that a substantial expansion in capacity was desirable but at the same time were greatly concerned lest undue speed result in dilution of quality or unevenness of output. In view of Google's obvious talent for rapid growth, he asked whether Mr. Shriram saw a way to accomplish the expansion quickly. He also requested Mr. Ahluwalia's opinion on the issue of liberalization in the educational sector, which in his own view might offer an alternative solution.

Mr. Ahluwalia answered that there was no doubt of the need for a massive expansion of India's education sector, adding that this major issue has been raised explicitly in the discussion document recently sent by the Planning Commission to the states and is to be debated over the next year or so. In the Commission's view, the expansion of existing publicly funded institutions, such as the ITTs, should be explored and the sector opened to private investment. There are no policy bars to privately funded, nonprofit institutions that, in addition to covering their costs, might generate a surplus that could then be reinvested. However, the deputy chairman had heard it said that regulations have kept the liberalization from being truly effective, an opinion he was inclined to accept. The next step is to look at the possibility of changing those policy elements that people wishing to set up new institutions had found restrictive. Not all the engineering schools or even the business schools in India are publicly funded, so the extent to which the restrictions were a problem and whether irritants might be removed by clarifying policy could be learned from these institutions.

Applying Google's Innovation Model to Agriculture

Alok Sinha, a professor of mechanical engineering at Penn State University and an alumnus of IIT Delhi, asked whether Mr. Shriram's innovation models would be applicable to the challenges facing India in agriculture, social infrastructure, and other fields, as outlined by Mr. Ahluwalia. He specifically asked for comments from Mr. Shriram on the potential impact of flat management structure on innovation in those areas.

Mr. Shriram demurred when it came to the application of Google's practices in other contexts. He explained that he and his colleagues at the company "thrive in an environment of constructive chaos." He cautioned, however, that the chaos must be kept "within reason," which is accomplished by ensuring that project teams are small and that results are measurable and receive close attention. A weekly review of the company's "Top 100 Projects" makes progress visible "on a dashboard" at the CEO and management levels. He speculated that it would be somewhat harder to achieve this on a much broader basis, as represented by agriculture or by India's other main areas of challenge.

Mr. Ahluwalia moved the audience to laughter with the observation that "agriculture, as an industry, actually has the flattest management structure: It's a lot of private farms, and they're all reporting to themselves." Then, addressing what studies have revealed as a "huge knowledge deficit" in the sector, he stated that Indian farmers could in fact be provided much better access to information than has been done traditionally. Moreover, the importance of doing this was increasing as farming moves toward objectives beyond simply growing enough food. "The first Green Revolution was about wheat, then rice," he recalled. "Today's Green Revolution is going to be about farmers growing a multiplicity of products."

High-Tech Solutions for Farmers and Fishers

Now providing farmers with electronic access to information were India's Krishi Vigyan Kendras, or Farm Science Centers, one of which the minister had visited one month before. This center, which serves a district of roughly 1 million people, hosts a Web site offering information about the kinds of crops growing in the area; diseases to which particular crops are vulnerable, with pictures to help identify the diseases; and recommendations formulated according to local conditions.[4] Those sending in questions by e-mail would receive answers based on consultation with the agricultural research university in the area. Weather information, the single most important kind of information that Indian farmers need, is also posted on the Web site; it comes not only from the Indian Meteorological Service but also from some U.S. forecasting centers. The extent to which farmers are making use of these services remains a question, but the deputy chairman pointed to the likelihood that connectivity improvements, both those in progress and those planned, would increase their opportunity to do so. Noting that Google had logged many hits of late from people in India looking for the truth behind the *Da Vinci Code*, he expressed his hope that Indian farmers would soon be retrieving information useful to them from such information storage and retrieval devices.

---

[4]The Krishi Vigyan Kendra's Web site can be accessed at <*http://aimlab.aces.uiuc.edu//diglib/india/kvk_index.htm*>.

This use of technology to reduce the knowledge deficit constituted a sea change, according to Mr. Ahluwalia, who said he could offer any number of examples but would limit himself to one more. In Pondicherry, there had been strong resistance to modernizing fishing boats on the grounds that it would disrupt traditional livelihoods. Now, however, the Indian Remote Sensing Satellite provides information on temperature conditions in the Indian Ocean—and, from there, to map onto where the fish were likely to be. As a result, fishers there underwent a "sudden change in attitude." They next realized that in order for them to capitalize on this information, they would have to invest in better boats. "Major changes are taking place in the lives of ordinary people, driven by what is otherwise very high technology," he remarked. And while the problem of access was as yet unresolved, the importance of finding a solution had become obvious.

Establishing U.S. Campuses in India

Vasant Telang, an associate provost at Howard University, asked about the panelists' reaction to the idea of establishing campuses of U.S. universities in India at which Indian students would follow the U.S. curriculum and receive a U.S. degree. Noting that his own academic background was in pharmacy, he observed that the United States is facing a tremendous shortage of pharmacists. Conversations with facilitating agencies, he said, had brought forth no objection.

Mr. Ahluwalia replied that although regulatory approval would be required to establish a degree-granting institution in India that recruited Indian students, it was unclear that merely locating a university campus in India would require any permissions. If what Dr. Telang had in mind was that an Indian student, after being accepted by a university in the United States, could then opt to go to a campus of that university that was located in India, then his guess was that no regulation would apply because that would be equivalent to renting space or putting up a building. However, since no such proposals had yet come in, he cautioned, a definitive answer called for further checking.

Opening Collaboration to Entrepreneurs

Anand Das, a former Silicon Valley engineer who spent the previous five years in the employ of the U.S. government, asked how U.S.-based entrepreneurs could participate in programs associated with the bi-national cooperation being discussed. He noted that some agencies, in view of the specialized nature of federal procurement processes, maintained program offices that dealt specifically with entrepreneurs.

Dr. McCormick stated that, of the numerous forums trying to increase collaboration between entrepreneurial communities in the two countries, he is most familiar personally with the U.S.–India High Technology Cooperation Group. Although it had involved mainly large companies at the outset, this group's focus has been evolving over the previous year to include start-ups, primarily in the life sciences and biotechnology. Its agenda for the next year and subsequent years emphasizes bringing in entrepreneurs in other areas: venture capitalists, U.S. private equity firms seeking opportunities in India, and similar Indian firms seeking opportunities in the United States. That forum, in which hundreds of business people from both countries had been engaged, had proved a constructive one.

Dr. McCormick concluded the session by saying that the number of questions and the line of questioners that remained at the microphones were indications of its success. Noting that the panelists would remain in the auditorium during the break to take further questions, he asked the audience to join him in expressing appreciation for what he termed a terrific discussion.

Synergies and Gaps in National
and Regional Development Strategies

Moderator:
Praful Patel
The World Bank

Dr. Wessner welcomed the panel and its chairman, Praful Patel, the World Bank's vice president for South Asia. Before ceding the podium to Dr. Patel, Dr. Wessner praised him for devoting his life not only to studying the questions before the panelists but, "even better, to doing something about advancing the development agenda in India and elsewhere."

Dr. Patel greeted the audience with the observation that the opening panel, in its focus on the bigger picture, had provided extremely good background for subsequent sessions. He observed that the theme of this panel, "Synergies and Gaps in National and Regional Development Strategies" is very closely aligned to the ongoing work of the World Bank and the government of India. Given the exciting progress on this topic, he said that he was pleased with the prospect that this panel would add to his knowledge.

Dr. Patel pointed to the presence among the panelists of Carl Dahlman, the lead author of *India and the Knowledge Economy*, a report published by the World Bank in 2005 that was of consequence for the day's discussions,[5] and to the presence in the audience of R. A. Mashelkar, who was to speak on a later panel. A workshop to be hosted on July 4 by India's Council of Scientific and

---

[5]Carl Dahlman and Anuja Utz, *India and the Knowledge Economy: Leveraging Strengths and Opportunities*, Washington D.C.: World Bank, 2005.

Industrial Research (CSIR) under Dr. Mashelkar's leadership was to take up the initial findings of the Bank's report.

Introducing all three members of the panel in order of appearance, he started with T. S. R. Subramanian, who has retired following a distinguished, 37-year career with the government of India during which he had held its highest civil-service position, Cabinet Secretary to the government of India. Then would come Dr. Dahlman, currently Luce Professor of International Affairs and Information Technology at Georgetown University, who had distinguished himself during more than 25 years at the World Bank, where he and Dr. Patel had been colleagues. The final presenter was to be Surinder Kapur, who was participating in the capacities of both chairman of the Confederation of Indian Industry's Mission for Innovation in Manufacturing and of founder chairman and managing director of the Sona Group. Dr. Patel assured the audience that what Dr. Kapur had to say about the group's activities would be very exciting and especially germane to the topic of how innovative technology's impact can be scaled up to reach large numbers of people on the ground.

Dr. Patel suggested that each of the speakers take no more than 15 to 20 minutes, so that time was certain to remain for questions from the floor. With that, he invited Mr. Subramanian to the podium.

BUILDING REGIONAL GROWTH:
ELEMENTS OF SUCCESSFUL STATE STRATEGIES

T. S. R. Subramanian
Government of India (retired)

Thanking Dr. Patel and greeting his fellow presenters and the audience, Mr. Subramanian said he would begin where Mr. Ahluwalia had left off by addressing areas for further action. The broad theme for the present panel could be the subject of numerous separate presentations, each of which might approach it from a different direction: by considering questions of national versus state government; urban versus rural; industrial, service, or agrarian policy; even regional and subregional development. Additionally, it could be addressed from the point of view of the entire framework of government policy, comprising policies of the central and state governments alike, or by asking whether the policies of the state governments were in conformity with those of the central government. He proposed to touch, however briefly, on some of these areas.

For the benefit of those lacking familiarity with the Indian system of governance, Mr. Subramanian noted that, as in the United States, the Indian Constitution clearly demarcates the powers and responsibilities of the central government from those of the state governments. Most of what touches the average citizen, including law and order, education, public health, and rural development, is in the purview of the states. Much of the reform to date had related to areas within the purview of the central government, such as monetary and fiscal policies.

Given that Mr. Ahluwalia had evoked India's successes, as exemplified by the high average annual growth rate achieved since the introduction of liberalization reforms, and because these very real successes are well documented, Mr. Subramanian said that he would not dwell on them. Instead, he would "zero in straightaway on the dark side not generally highlighted in the euphoria of India's great progress": the arena of human development, where the numbers, he asserted, were "terrible, to say the least."

After nearly 60 years of self-rule, more Indians are now below the poverty line—which Mr. Subramanian called "a euphemism for saying they're desperately poor, they don't know where the next meal is coming from"—than at independence. The divide between urban rich and rural poor has dramatically increased. The majority of the population in most urban areas, including metropolitan areas, consist of slum dwellers, the average being about 55 percent for the country as a whole—"slum dwellers," as he characterized them, "in the glittering cities."

That urban areas act as magnets despite the harsh conditions they offer the poor is indicative of the lack of job creation in the rural context, Mr. Subramanian suggested. Whereas primary education is universal and compulsory, only 16 percent of the population complete high school; just doubling that number would have a significant impact on the base for information technology (IT) development. Meanwhile, public health is in bad shape, with rural facilities "abysmal." The micronutrient level in the human diet is the same in India as in sub-Saharan Africa. Even at its current rate of growth, India would not catch up with Africa's current average until 2025. "I could go on," he told the audience.

Mr. Subramanian said his point was this: "Two vastly different Indias cannot coexist peacefully as a democracy for long." Everything relating to public health, education, and other "soft" areas falls within the purview of state governments. The winds of change have not yet reached these sectors, not because of lack of awareness but because of a lack of will arising from local political compulsions. The culture of corruption in state and local governance means that, as Rajiv Gandhi had not so jocularly estimated, only 15 percent of the funds earmarked for development actually reach their intended beneficiary.

Although every party to hold power has promised to reach out to the poor that comprise three-fourths of India's population, there has been no concerted, meaningful effort by the central or the state governments to change this picture. The unspoken presumption has been that all boats would be lifted according to "the trickle-down theory," but the efficacy of this "strategy" has not become evident—on the contrary. "In an era when information is easily transmitted to the remotest area, and awareness levels have grown, this contrast" between rich and poor, Mr. Subramanian warned, "is not sustainable for long."

But many state governments, which clearly would have to drive the next series of steps, are now reorienting their approach to rural development strategy.

With two-thirds of the population drawing their livelihood from agriculture, rapid agricultural growth remains the key to poverty alleviation. In the 1950s and 1960s, he recalled, few observers gave India any chance of becoming self-sufficient in agriculture. Indeed, the Club of Rome called the country a "basket case." The Green Revolution proved the pundits wrong. Though the work was driven locally, led by such people as C. Subramanian and M. S. Swaminathan, there was seminal technical assistance from the United States, and through the World Bank's intervention.

Cause for concern of late, however, is the deceleration in both production and factor productivity growth recorded in some of the major irrigated production systems, and many areas still do not have access to seasonal irrigation. Although public research and extension programs have played a major role in bringing about the Green Revolution, old methods relying on government servants to transfer technology clearly have severe limitations. The old extension systems are no longer relevant, and no effective new system is yet in place. The state agricultural universities and rural engineering colleges that were designed to support the agrarian sector are hopelessly out of date. New methods, practices, and techniques of delivering technology assistance via the private sector and voluntary agencies are urgently needed. A number of states are already working on these areas, with experiments under way.

Strong Indo–U.S. collaboration, if appropriately designed, could clearly foster a Second Green Revolution, in Mr. Subramanian's judgment. Strong commitment at the state level in India would also be necessary, however.

Until three years before, when Monsanto had introduced Bt cotton on 45,000 hectares, that variety had been unknown in India. Now Bt cotton is sown on 1.2 million hectares, accounting for 15 percent of the country's cotton cultivation, and it is expected to account for as much as 50 percent within two to three years. It is imperative to acknowledge, however, that both the pricing policies for Bt cotton and its environmental impact are the subject of "furious" debate in India. The astonishing support enjoyed by the new cotton varieties should be seen side by side with the fact that over 500 farmers commit suicide every year in each of the cotton-growing states. "This, of course, also relates to other aspects of rural reforms," he said, "like ushering in proper lending policies coupled with crop insurance systems."

So, there is a major and urgent potential in India for practical and meaningful reform. Although the main effort has to be made within the country, assistance of a critical nature—technical and technological—could be provided by appropriate U.S. agencies. America's principal gain from the first Green Revolution, which took place during the Cold War, related to the advancement of its geopolitical interests: entry into the Indian political, official, academic, and media zones, which had over time yielded benefit to the United States. With India's vast rural market potential, partnership in the agricultural sector today can yield rich economic dividends for all.

Interventions in that sector have to be carried out with caution, however. Contract farming is still a bad word in many rural circles, owing to local political sensibilities. "Similarly," Mr. Subramanian said, "it would be naïve to think that the mere entry of large retail produce-marketing chains from abroad would automatically bring joy to India's rural areas." Severe physical infrastructure bottlenecks have to be addressed before the entry of large corporations could add significant value through improving products or generating employment. There is no magic formula in such things.

Among the most pressing requirements for rural development are those found in the fields of energy and education, while those in the field of public health are likewise enormous. Power generation—and especially transmission and distribution to rural areas—is a realm in which the United States and other countries could clearly play an important role. Solar energy is another field offering potential. Opportunity is also ripe in rural India in computer learning, particularly at the lower levels, as new and improved systems for delivering primary and secondary education are of high priority. Significant experiments, some receiving technical and financial support from the United States, are already in progress.

Higher education requires urgent attention as well. Of the more than 500 engineering schools and 600 management institutes in India, only the Indian Institutes of Technology (IITs) and the Indian Institutes of Management are world class, while the rest—a majority—were greatly in need of improvement. According to the assessment by Sam Pitroda, a noted futurologist, India's requirement for high-quality institutions of higher learning in the coming decade stands at around 200. Most of such institutions were in the purview of state governments, many of which support the upgrading of existing institutions and the initiation of greenfield projects. This is clearly an area offering mutual benefit to India and the United States through the interchange of faculty at all levels.

Prerequisite for the successful implantation of the desired new processes, Mr. Subramanian declared, would be both a major drive to eradicate corruption at all levels and assurance of significant judicial recourse. "No doubt, India has an extremely well-developed and mature judicial structure," he commented, "but 'justice delayed is justice denied' is not currently the guiding principle."

The week before the conference, Mr. Subramanian had passed a brief holiday at a remote Himalayan village with a total population of about 1,000 living in 10 or so hamlets. In the central marketplace, international telephony and Internet were available and working well. There was even a tiny computer school with all of two computers teaching C++ and graphics. In fact, a high-speed multipurpose underground coaxial cable was being laid in that remote mountain area during his stay. However, while a few new service-sector jobs had been created, there had been no significant job creation in the diversified agrarian sector.

This small hamlet, in Mr. Subramanian's view, represents the opportunity and the challenge facing India. "The winds of change have come and can no more be resisted," he said. "Changes of one sort or the other will, willy-nilly, take place." In the absence of rapid economic development in India's rural areas, renewed questioning of the democratic process would take place, a fact well known to the nation's political decision makers. Therefore, the opportunity for rapid development was not to be missed.

The leadership at the state level has not only to recognize this, but to act. It is imperative to introduce new players, including the private sector and voluntary agencies, which could take on major responsibilities under appropriate safeguards. The bright side was that significant groundwork has already been laid and the remaining work could be accomplished with relative ease in a relatively short time. "All it needs," Mr. Subramanian concluded, "is a small dose of political will driven by the poor and uneducated Indian."

INDIA'S KNOWLEDGE ECONOMY IN A GLOBAL CONTEXT

Carl J. Dahlman
Georgetown University

Observing that much was at stake in the day's discussions, as they concerned two very large and important countries whose further collaboration would be of mutual benefit, Dr. Dahlman called it an honor and a pleasure to participate. He advised the audience that he had so many slides to present that he would go through a number of them quite quickly, emphasizing the key points.

He began by positing that the world was in the middle of what could be called a knowledge revolution. Much new technology was being created, some of it so rapidly that it was very important for countries to develop effective strategies to deal with it. In the case of India, with its "tremendous potential," this was particularly applicable.

To situate India on the global stage, Dr. Dahlman noted that it is home to 17 percent of the world's population and has the eleventh largest economy as measured by nominal exchange rates but the fourth largest in terms of purchasing power parity (PPP). Substituting PPP for nominal exchange rates as the yardstick causes India's share of world GDP to rise from 2 percent to 5.6 percent and thus reveals it as much larger than it might seem. India's economy is still relatively closed, however, accounting for only 1 percent of world trade, and it faces strong competition from other countries, China in particular.

Comparing innovation indicators for India and China, he stated that the latter has integrated much more rapidly into the global market: Trade in manufactures accounts for 51.3 percent of China's GDP in 2004 against 13.5 percent of India's, a "very stark" difference. Similarly, 27.1 percent of China's manufactured exports fall into the high-tech category, with India's corresponding figure reaching only 4.8 percent.

Other indicators point in the same direction. China has more than seven times India's foreign direct investment as a percentage of GDP—5 percent vs.

0.7 percent—and a like margin, 810,525 to 111,528, in number of researchers engaged in R&D in 2002. Additionally, China is acquiring $2.75 worth of technology through formal transfer for each member of its population versus India's 40 cents per person, said Dr. Dahlman. And China spends 1.4 percent of GDP on research and development, whereas India's share remains at around 0.8 percent. Finally, China is producing about twice the number of scientific and technological journal articles as India, 16.5 vs. 10.73 per million population in 2001, and it was granted 597 U.S. patents in 2004 to India's 374.

In producing *India and the Knowledge Economy* for the World Bank, Dr. Dahlman and his coauthor developed a method for assessing how national economies are faring in this new environment based on four of their components, or "pillars": economic and institutional regime (EIR), education, information infrastructure, and innovation. A country's EIR governs its "incentives for the efficient creation, dissemination, and use of existing knowledge," thereby determining the degree to which its economy can restructure itself to take advantage of new opportunities. Education and skills constitute an extremely important topic and one to which Dr. Dahlman would return. The information infrastructure reduces transaction costs, allowing much to be done far more efficiently, and takes knowledge out to the rural sectors, an issue raised by Mr. Ahluwalia and Mr. Subramanian. However, the main focus of his talk was innovation, which covers not only a country's domestic R&D, as important as that is, but also—and equally important—the ways in which an economy taps into existing global knowledge: via foreign investment, trade, technology transfer, education abroad, the Internet, and technical publications.

The report's authors also developed a scoring methodology based on about 20 indicators for each of the four pillars, which he demonstrated by projecting a graphical comparison of India, China, and the United States that used a reduced number of variables (Figure 1).

A country's place in the rank ordering is indicated by its distance from the circumference of a circle, and, as might have been expected, the United States is closest to the circle in the chart shown. Running through the variables clockwise from the top of the circle, Dr. Dahlman broke them into five categories:

- **Performance**—GDP growth and the Human Development Index;
- **Economic and Institutional Regime**—tariff and nontariff barriers (which he designated a measure of competition), regulatory quality, and the rule of law;
- **Innovation**—researchers in R&D, scientific and technical journal articles, and patent applications granted by the U.S. Patent and Trademark Office;
- **Education**—adult literacy rate, secondary enrollment rate, and tertiary enrollment rate;
- **Information Infrastructure**—fixed and mobile telephone lines per 1,000 people, computers per 1,000 people, and Internet users per 10,000 people.

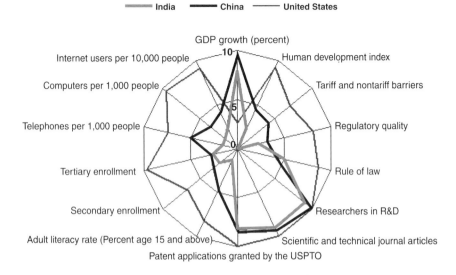

FIGURE 1 Comparison of China, India, and the United States.
SOURCE: World Bank Institute, KAM 2006, *<http://www.worldbank.org/kam>*.

A similar chart depicting the status of this "very serious competition" 25 years before, he said, would have shown India ahead of China in just about every variable, whereas China has now surpassed India in everything but democracy and some aspects of the rule of law.

He then projected a graph representing spatially the relative positions of numerous countries in the global knowledge economy (Figure 2).

Plotted according to the horizontal axis was a country's 1995 position on all the indicators used in the circular chart in Figure 1 minus GDP growth and the Human Development Index; the vertical axis represents its position in 2003–2004. Position along the diagonal indicated level of development, with the more developed countries situated in the upper-right quadrant, the developing countries toward the middle or in the lower left; India was to be found in the sixth decile. Position relative to the diagonal indicated whether a country's performance had been better in 1995 or in 2003–2004, with those whose performance had improved rising above the line and those whose performance had deteriorated falling below it. India has fallen back a little, which he said should probably be seen as cause for concern. The main reasons for its decline are related to two of the four "pillars," EIR and education; when it came to information infrastructure and innovation, India had shown improvement.

This was not to deny that India has registered tremendous accomplishments over the previous two decades, as Mr. Ahluwalia had outlined. The increase in its

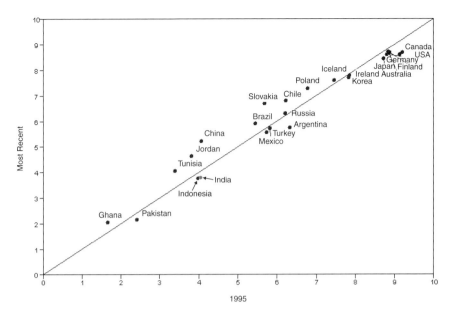

FIGURE 2 India in the global knowledge economy.
SOURCE: World Bank Institute, KAM 2006, <*http://www.worldbank.org/kam*>.

annual average rate of GDP growth—from 6 percent over the 1990s to 6.2 percent from 2000 to 2004 and then 8 percent over the previous three years—is extremely impressive. Moreover, the country has potential for continued acceleration, which in any case is needed to provide opportunities for its fast-growing population and even faster-growing workforce.

Dr. Dahlman showed a graph (Figure 3) with projections of a dozen major economies' real per capita GDP growth between 2004 and 2015 in terms of purchasing power parity; these projections were based on the economies' records of growth over the previous decade.

Plotted according to the vertical axis was GDP in trillions of dollars, with the horizontal access representing time. The chart shows China surpassing the United States by around 2013, while India, already the world's fourth-largest economy in PPP, is to move past Japan into third place by around 2008. The actual and projected speed of growth of China and India, about three times the world average, is evidence that both economies will become increasingly powerful. For this reason, it is extremely important to keep track of what India is doing, but it also bore remembering that India faces significant competition, particularly from China.

He then recited India's many fundamental strengths: its very large domestic market, young and growing population, critical mass of educated people, very strong R&D infrastructure, and strong science and engineering capabilities cen-

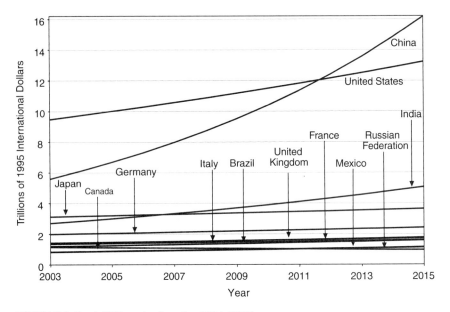

FIGURE 3 Real GDP projections for 2004–2015.
SOURCE: Projections based on data in the WDI database. World Bank, World Development Indicators 2006, Washington, D.C.: World Bank, 2006.

tered in areas such as chemicals, pharmaceuticals, and software. In addition, India:

- is becoming the world center for many digital services, a location where "anything that can be offshored" can be done very cost-effectively;
- is becoming a center for contract innovation for multinational companies, which have established around 400 R&D centers in India to tap its most valuable human resource, its scientists and engineers;
- enjoys a network based in the worldwide Indian diaspora, strongly represented in the United States, which could prove an excellent source of everything from information and advice to access to markets, technology, and financing as India's activities increased in sophistication;
- boasts very deep financial markets, far better than China's; and
- is beginning to strengthen its export orientation and to seek strategic alliances.

As for the challenges India faced, Dr. Dahlman said that its very large population—nearly 1.1 billion at present and estimated to grow at an annual rate of 1.4 percent in the period 2004–2020—needs to be "skilled up." India's adult population averages 4.8 years of schooling, and its illiteracy rates are 52 percent for women and 27 percent for men. This very low average level of educational attainment is India's "biggest Achilles heel" in a world economy where the capacity to absorb and use modern knowledge is so important. China, beginning at a lower level 15 years before, has improved its average education level to eight years, whereas for India this remains a gigantic challenge. Other challenges stem from the fact that India's economy is overregulated, that it has very poor physical infrastructure, and that it is competing in a very fast-paced, global environment where speed is of the essence.

Dr. Dahlman acknowledged Mr. Subramanian's contention that there are two Indias: the high-tech India with which many abroad were familiar and that many in the audience represented, and a very poor India. "It's an extreme dual economy," he said, "and it is very important to see how we can leverage traditional knowledge and modern knowledge to address the needs of a very large population that is still far behind." For, at about $620, India's per capita income remains low.

Nonetheless, India now has a very special window of opportunity during which it could undertake key reforms and leverage its strengths to improve its competitiveness and the well-being of its people. However, even if all the critical ingredients were present, the country has to marshal them well, and time is of the essence. Policy choices would matter: Dr. Dahlman and his World Bank coauthor had made four projections of India's per capita income through 2020, assuming a constant rate of investment but varying the level of total factor productivity (TFP) growth. They discovered that TFP could make a difference of over 50 percent in per capita income within 10 years. In what might serve as a guide for Indian decision makers, he offered what he called "key areas for action," organized according to the four pillars of the knowledge economy cited in the report.

Improving the Economic and Institutional Regime

Dr. Dahlman called this step a fundamental one in that it would set the entire context for change. This involves:

- reducing bureaucracy for the entry and exit of firms;
- upgrading physical infrastructure, with a focus on power reliability and the efficiency of roads, seaports, and airports;
- easing restrictions on the hiring and firing of workers;
- reducing tariff and nontariff barriers, because the Indian economy remains among the world's most closed, whereas China, having reduced nontariff barriers since joining the World Trade Organization, has become globally very integrated in pure trade structure;
- encouraging more foreign direct investment and increasing linkages with the rest of the economy;
- strengthening intellectual property rights and their enforcement; and

• improving governance and using information and communication technology to improve transparency, something in which India was making considerable progress, although more remains to be done.

Strengthening Education

This would eliminate a basic constraint on India's advancement. The country provides a small elite with excellent education, but it has quality problems, even in higher education, outside the Indian Institutes of Technology and Science. Important measures would include:

• expanding the quality of primary and secondary education;
• raising both the quality of higher education and the number of teachers, the latter being held down by regulatory obstacles even though salaries were beginning to go up very rapidly;
• embracing the contribution of private providers of education and training by lowering bureaucratic hurdles, among which is an accreditation system that has frustrated many large private universities;
• developing partnerships between academia and industry to increase the universities' awareness of the skills required to create the knowledge workers necessary for economic progress;
• using information and communications technologies to meet the double goal of expanding access to education and improving its quality—which would involve little more than realizing the tremendous potential India has, given its lead in the sector, to reach out to its very large population; and
• investing in flexible, cost-effective job training programs that can adapt quickly to new skill demands. With the half-life of knowledge growing ever shorter, a massive lifelong learning system allowing those who had left the formal education sector to upgrade their skills was critical. High-tech firms in India, one-third of whose employees are getting more education at any given time, are leading the way, but their efforts have to be expanded in the interest of increased competitiveness.

Leveraging Information and Communications Technologies (ICT)

This step begins with boosting ICT penetration throughout the country, which would require:

• resolving some regulatory issues, as well as lowering some import costs, the latter by reducing or rationalizing tariffs on hardware and software;
• massively enhancing ICT literacy, perhaps through programs such as those South Korea had undertaken to build skills enabling its people to use the Internet and other information and communication technologies;

• exploiting ICT as a competitive tool by using it to increase the efficiency of production and marketing through enhancing supply-chain management and logistics, as well as to improve the delivery of government services—a promising area because India has firms capable of producing important innovations;
• moving up the value chain in information technology; and
• providing suitable incentives to promote information-technology applications for the domestic economy, including local language content and applications.

Strengthening Innovation

The "pillar" he had saved for last was the one Dr. Dahlman most wanted to emphasize. He pinpointed areas in which India should make a greater effort:

• **Tapping global knowledge:** Its current low share of total world spending on R&D—less than 1 percent or, in PPP terms, about 2.5 percent—indicates that India needs to tap more effectively into the knowledge available beyond its borders. China, in contrast, is doing this very well by bringing in foreign investment, a source not only of capital and capital goods but also of management skills, technology, and access to markets.
• **Attracting foreign direct investment:** As indicated, India has a long way to go here despite its liberalization, as many sectors remain closed. Furthermore, many foreign investors are being discouraged by the country's cumbersome bureaucratic structures, corruption, and poor physical infrastructure. It is because of the constraints involved in the physical movement of products in and out that the country's biggest growth has been in products and services that can be transported digitally.
• **Making use of its diaspora:** Taiwan, Korea, and China have developed effective networks to take advantage of the expertise, skills, and market information of its nationals living outside the country. India is just beginning to do the same, and, Dr. Dahlman observed, many in his audience would likely be able to increase their contribution to its progress.
• **Improving the efficiency of public R&D:** India's total R&D spending comes to only 0.8 percent of GDP, of which around 70 percent is in the government sector. In turn, around 70 percent of that goes to big, mission-oriented programs in defense, oceans, and space, leaving projects more oriented toward the basic needs of the economy—in agriculture, industry, and health—with quite a small share. Dr. Dahlman raised the possibility of redeploying resources to those areas, as well as of increasing the efficiency with which resources were used. Praising Dr. Mashelkar for having done "a remarkable job" in changing the incentive structure of the CSIR laboratories to orient them more toward such needs, he suggested that Dr. Mashelkar's approach be generalized so that it reached more of the public infrastructure. Also important was moni-

toring how effectively resources are used, and once monitoring systems have been improved, looking at the government's share of R&D spending would be critical.

• **Motivating private R&D investment:** The private sector accounts for only 23 percent of India's total R&D spending, about one-third the level of traditionally industrialized countries. In China, the corresponding figure has reached 65 percent. And, much of this R&D is being done by foreign multinationals, which are performing R&D at 400 labs located in India, proof of the presence in India of a critical mass of scientists and engineers. Why are domestic firms not doing more? Thought should be given to the kinds of programs that might be put into place to encourage them.

• **Shoring up university–industry programs:** Matching grants might be one way of fostering interaction between the public and productive sectors.

• **Developing supporting institutions:** Dr. Dahlman named as very important to a competitive economy, and thus worthy of support, science and technology parks and incubators; early-stage financing and venture capital; and metrology, standards, and quality control.

• **Enforcing intellectual property rights:** Taking steps to protect the fruits of innovation would create confidence among both domestic and foreign researchers.

• **Nurturing grassroots innovation:** In need of stronger support was India's already very extensive program of grassroots innovation, which addressed many of the needs of those left out of the modern economy.

• **Bolstering formal innovation systems:** The advanced technology coming out of both public and private innovation systems could help improve broad social and economic conditions, an example being Mr. Ahluwalia's account of the application of remote-sensing satellites in fishing. "More could be done to find ways of harnessing modern knowledge to meet the very pressing needs of a very large poor population," Dr. Dahlman said.

• **Broadening science and engineering education:** India's handful of elite institutions, seven IITs plus the Indian Institute of Science, produced only 7,000 science and engineering graduates annually for an economy of more than 1 billion people. This supply shortage needed urgent attention.

Dr. Dahlman concluded by addressing what he called the "tremendous" opportunities that existed for U.S.–India cooperation. Increases are to be envisioned for foreign direct investment going in both directions, for trade in goods and services, and for strategic alliances—notably in pharmaceuticals, software, auto parts, and chemicals, but in many other sectors as well. Cooperation in education and training also holds out great promise, particularly for unleashing India's high human capital potential. Finally, ample occasion will arise for the two countries to collaborate within the framework of international programs in such areas as

energy and environment. There will be great mutual benefit in increased cooperation, he concluded.

MANUFACTURING INNOVATION AS AN
ENGINE FOR INDIA'S GROWTH

Surinder Kapur
Sona Group

Expressing appreciation for the invitation to speak and greeting Minister
Sibal and Dr. Mashelkar, who were in the audience, Dr. Kapur announced that he
would devote his attention less to India's achievements than to gaps that remained
and the ways in which they are being addressed.

Dr. Kapur introduced himself as chairman of the Mission for Innovation in
Manufacturing of the Confederation of Indian Industry (CII). CII is a 110-year-
old body that represents over 5,000 members organizations; it is, he said, truly
the voice of Indian industry. Earlier in 2006, CII established the Innovation in
Manufacturing Mission in the belief that it would fall to manufacturing to create the employment required for India's further economic growth. With India
famous the world over for information technology (IT) and IT-enabled services,
few recognize that manufactures account for 75 percent of the country's exports
even though the sector represents only 20 percent of its GDP. Because of the
manufacturing sector's role in providing jobs, examining it closely to see how
its performance might be improved is vital.

Dr. Kapur noted with contentment that, only days before, IBM had announced a commitment to invest $4 billion in India. Similarly, General Electric
had announced its intention to increase its Indian revenue from $1 billion to $8
billion. "India is, in my view, rocking," he exclaimed. "I think there are great
opportunities."

Beginning his discussion of innovation, Dr. Kapur recalled a statement by
S. Ramadorai of Tata Consultancy Services (TCS), whom he described as India's
leading CEO. Ramadorai had said of TCS's IT-enabled services: "Cost helped
us get a foot in the door, quality opened it a little bit more, and now we need innovation to open it all the way."

CII recognizes, according to Dr. Kapur, that the people-cost arbitrage upon
which India has been relying can keep it competitive in the international environment only in the short term. While contract manufacturing currently provides
some growth, India needs to reposition itself from low-cost manufacturer and
service provider to creative product developer. The Mission for Innovation in
Manufacturing is CII's contribution to this effort.

The mission's objective is to serve as a facilitator to help 100 Indian companies develop as leaders in innovation and product development over the next three

years. Although this target might not seem very ambitious, CII believes, he said, that these leader companies can create a much-needed ripple effect throughout the country. To help promote the concept of innovation and its associated processes among Indian manufacturers, the organization has proposed to open institutional framework cooperation with such global authorities on innovation processes as Clayton Christensen of Harvard University, Deming Prize winner Shoji Shiba, and Patrick Whitney, director of the Institute of Design at Chicago's Illinois Institute of Technology. It is also partnering with India's own National Manufacturing Competitiveness Council (NMCC), formed by the prime minister in 2005 to develop a strategy for enhancing the country's manufacturing.

CII's main goal is to catalyze the desired transformation through changing the attitudes of those at the top of manufacturing companies. To find the 100 companies that will emerge as its leaders, the mission is identifying 500 companies with which to begin working. These firms would be introduced to the mission's roadmap and have their processes benchmarked to help them adopt innovation as their strategy for growth.

As evidence of the prevalence of advanced skills' in India, Dr. Kapur posted a list of familiar names in U.S. research that are operating R&D centers in India: Bell Labs, Cognizant Technologies, Enercon, Exxon, GE Industrial Systems, GE Medical Systems, IBM, Intel, Lucent, Microsoft, Motorola, National Instruments, Oracle, SeaGate, Texas Instruments, and Xytel. "We believe there is a lot of culture that Indian manufacturing companies need to take advantage of," he observed.

To demonstrate that innovation is indeed appropriate as the next phase in the country's development, he provided an outline of the quality movement in which CII has been engaged for decades. Beginning in the 1960s, the organization trained manufacturers in the same process control of associated with Edwards Deming and that had led Japanese manufacturers to such great heights. Practices associated with it—among them gap analysis, corrective countermeasures, and preventive countermeasures—are now in use throughout the world.

From teaching process control, CII had gone on to imparting the tools for continuous improvement. He stated proudly that a strong culture of continuous improvement is developing within Indian manufacturing companies. Not long before, Dr. Kapur had the privilege, as an NMCC member, of presenting India's prime minister with two CD-ROMs on which were recorded 1,000 *kaizens*, or improvements, accomplished by Indian companies. He was currently in the process of putting together 100,000 *kaizens* for the prime minister to deliver to the nation as a way of emphasizing that the CII's unique method of teaching improvement represented knowledge that belonged to the nation rather than being contained in one particular organization.

Behind CII's claim of uniqueness lies the fact that India is the first country in the world to have used clustering in pursuit of quality. Formed into 32 clusters were some 270 noncompeting companies too small to afford the high-cost consultants that Dr. Kapur called "really necessary" to learning quality-improvement methods and that the CII itself invited from the United States, Japan, or Europe. He characterized the attitude within the clusters as "one of give and take: 'I learn from you and you learn from me so that we don't have to relearn the same things over and over again.'" Clustering had turned into movement that was upgrading quality across India, not only in advanced manufacturing but also in the leather and tea industries.

The CII had also made efforts in policy deployment with management for objectives, a powerful tool for improving results by ensuring that everyone within an organization is in alignment. Dr. Kapur proposed to share the experience of his own firm, the automotive component manufacturer Sona Group, a 16-year-old "boutique global company" with seven plants in India and minority investments in France, Brazil, and the Czech Republic that was posting about $200 million in sales annually. The firm had been able to make "quantum" changes within a short time, which he attributed to its being part of a cluster and to the particular set of tools it employed.

Sona's focus in its Total Quality Management activities, which began around 1998, is on skill building. In the three years ending in 2003, on a per-million basis the company has reduced its in-house rejections from 23,000 to 876, its supplier rejections from 49,500 to 996, and its customer returns from 1,579 to 112—and it is currently showing far better numbers than it had three years earlier. One illustration is its inventory turnover ratio: After jumping from 7.5 to 32 in the three years ending in 2003, it is running at close to 55, a number that included imported inventory, which in Sona's case was very high.

Productivity improved over the same three-year period, with gross sales per employee rising from 1.6 million rupees to 5.9 million rupees; currently, the firm was operating at the level of 7 million rupees. Indices of morale are his favorite, Dr. Kapur said, because they demonstrate "the people involvement we were able to generate within three years in this cluster program": Accidents have been all but eliminated, and absenteeism has dropped from 11.3 to 6 percent over the three years as training on a per-employee per-year basis has grown from 20 hours to 57 hours, with another 12 hours per employee per year added since 2003. Moreover, suggestions per employee per year has climbed from 2 to 20 in the three-year period and subsequently increased to 27.

Breaking through to innovation would be, in CII's view, the next step for India's manufacturing industry. In 2004, it established a new cluster under the guidance of Professor Shiba, who came every three months to spend three weeks working with all of the CII companies together in what was termed a "learning community." Dr. Kapur outlined the results that the program's four original participants have been able to obtain in six months with the use of Professor Shiba's methodology—which he calls "Swim with the Fish"—for creating an ambidextrous organization and an organizational architecture in which ideas are not killed:

- **TechNova**, holder of 85 percent of India's market for printing plates despite competition from Kodak and others, was able within half a year to create and roll out a new process, PolyJet BT, for wide-format inkjet printer imaging.
- **Sona Koyo**, a steering-systems company in Dr. Kapur's group, focused on skill development and improved productivity on a particular line by 30 percent within 20 days. It was also able to develop a new electronic power-steering product for the agricultural sector whose implementation it was currently discussing with a number of U.S. companies.
- **UCAL**, which manufactures fuel injection equipment in the south of India, developed a concept of "factories within the factory" in which even production workers deal with the customer. It currently had four factories within its main plant.
- **Brakes India Foundry** scored a breakthrough in environmental management, becoming a zero-discharge company—and, in the process, bringing a 4 percent saving down to its bottom line.

As a result of these innovation experiences, the four companies were themselves expanding, and more and more companies are signing up to work with Professor Shiba through CII's program.

There were, in fact, numerous signs that India is approaching world-class status in manufacturing as it prepares to ramp up job creation in the sector. Among them, it boasts (at last count) 16 Deming Award winners, 95 Total Productive Maintenance Awards from the Japan Institute of Plant Maintenance, and a Japan Gold Medal. Although, according to Dr. Kapur, the country is "in a great situation" demographically, poised to become the world's youngest nation over the next decade, it needs to accommodate between 8 million and 10 million new job seekers every year. Moreover, he acknowledged, its manufacturing and innovation strategy, geared of necessity to the goal of employment generation, faces a number of challenges. At the same time that India expands production capacity, both through foreign direct investment and domestic companies' enlarging their manufacturing base, it is also saddled with archaic labor laws that could not be changed overnight. These labor laws, pending reform, might prove an obstacle to cooperation with U.S. institutions.

For this reason, the CII was very pleased at both the formation of the National Manufacturing Competitiveness Council and its development of a strategy for increasing manufacturing's share of GDP to 25 percent from the current 17 percent. Posting a schematic representation of the strategy (Figure 4), he called attention to the fact that the prime minister was personally heading a mechanism for monitoring and measuring performance, the Committee on Growth and Competitiveness of Manufacturing. This was significant not only in demonstrating high-level commitment but also in making certain that, in cases where an intervention by the prime minister's office was required, it would cut across the bureaucracy from ministry to ministry. Because India needs to create demand

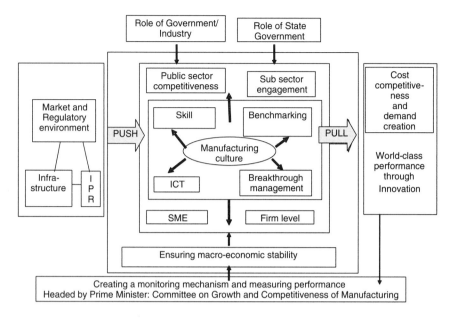

FIGURE 4 NMCC's manufacturing strategy

through cost competitiveness and to achieve world-class performance through innovation, both the national and the state governments have an important role to play: making the changes that support the environment required. "I'm certain that the manufacturing culture which is at the core of this will come about with the programs that we have put together and the strategies that the National Manufacturing Competitiveness Council has developed," he declared.

CII's contribution to this endeavor is to mount an awareness campaign to be kicked off by a visit to India of Christensen, Whitney, and others the following month. Dr. Kapur appealed to the audience for its support at a moment when Indian companies need training in the tools and techniques of innovation, since innovation, he observed, "doesn't just happen." In its initial year, the CII's Innovation Mission Plan would target 10 to 20 companies in each of five focus sectors: machine tools, automobiles and auto components, electrical equipment, chemicals, and leather. A number of skill-building tools would be employed in training the companies: Christensen's *Disruptive Technologies*, concept engineering, TRIZ, the Strategy Canvas as put forward in *Blue Ocean Strategy*, Whitney's technique for creating concepts, and breakthrough management techniques.

CII plans to continue forming high-tech clusters, in view of their effectiveness to date, and to bundle its programs to work systematically as part of a comprehensive approach. Coming in for particular attention would be small and mid-sized enterprises (SMEs), which, since they constitute "the real base of manufacturing," are needed to contribute to the sector's growth. CII, he said, is in the final stages of setting up an institute in India that, with the participation of Professor Christensen, would work with its SMEs; because small firms could seldom afford such prominent consultants, thought was being given to devising e-learning tools to assist them. "These learning communities are really the ones that are going to diffuse the competencies required for the country," Dr. Kapur said.

Noting that there were various kinds of cooperation that the CII might forge with institutions in the United States, Dr. Kapur expressed particular regard for U.S. institutions of higher learning: "I am so glad that so many people from academia are here," he said, "because that is really what is required." In light of India's strong motivation to generate jobs in manufacturing and of the increased attention paid the sector's problems and opportunities in recent years, he predicted: "We will create winners."

DISCUSSION

Thanking all three speakers for making rich presentations in such a short time, Dr. Patel remarked that their discussion of innovation had taken place in the larger context of India's development agenda. Mr. Subramanian's observation that some farmers were resorting to "the extreme measure of suicide," as Dr. Patel put it, serves as a reminder of the importance of delivering the results of innovation to all levels of India's diverse population; he had also mentioned the need to address corruption and judicial reform. Dr. Dahlman had talked about how to implement the agenda for India's near future in the context of what he had referred to as "two Indias." And CII's mission, as related by Dr. Kapur, which would allow the scaling up of manufacturing thanks to the ideas about innovation that he had described, also had a larger relevance for India. So, even though the talks had to a certain degree gone over familiar ground, they had been very interesting and had contained some new ideas.

Dr. Patel, opening the floor to the audience, said that he would take three questions at a time, and then ask members of the panel to address them.

Charles Wessner of the STEP Board complimented Dr. Kapur on his exposition of the importance to India of improving manufacturing, but he sounded a note of caution regarding the speaker's focus on collaborating with U.S. institutions of higher learning. The involvement of academics in the work of the National Academies, to which they frequently contribute, sometimes poses obstacles to obtaining policy-oriented results. Dr. Patel's colleagues at the World Bank, suggested Dr. Wessner, have excellent experience in providing practical policy advice.

Dr. Wessner then observed that, although Dr. Kapur had made reference to the importance of SMEs, he had spoken neither about how to encourage them nor about linkages between universities and small business. He noted the treatment that SMEs receive from the U.S. Council on Competitiveness: That organization speaks a great deal about the importance of small business but focuses most of its recommendations on the corporations and universities that are its main stakeholders. Common to many countries is an attitude that Dr. Wessner summed up thusly: "Small business is important; now let's talk about big business and universities." His question was whether India had programs at either the national or state level designed to encourage technology transfer from universities to SMEs or to provide the early-stage financing for entrepreneurs.

Dr. Sinha of Penn State, also addressing Dr. Kapur, noted the existence of a report on manufacturing by the National Academy of Engineering in which the term "customerization" is used.[6] This appeared to be the manufacturing-sector equivalent of tailoring products to the needs of individual customers and thus to suggest an economy or a manufacturing industry without mass production. He wished to know whether CII had taken such a concept into account in its manufacturing strategy.

Ajay Kalotra said the company he directed, International Business & Technical Consultants, Inc. (IBTCI), is active in the area of development and therefore serves as a consultant to the World Bank and other institutions. He noted that throughout the world, the biggest challenge is trying to figure out how to work on macroeconomic, microeconomic, and enterprise reforms all at once. Given what India is facing, there is no doubt that these reforms need to be implemented simultaneously, and quickly as well. Although Indians have every right to be proud of what they have achieved, the remaining challenges are such that all the progress CII might make within 100 companies would not amount a drop in the ocean for India, nor would bringing in three professors change the course of its history.

Mr. Kalotra had found two further causes for surprise in Dr. Kapur's remarks: that CII was addressing manufacturing exclusively when service industries were so important and that clustering was being fostered among noncompeting firms for the purpose of regional economic development rather than among competing firms so that they could combine to take on international markets. Noting that he had been away from India for 25 years after having worked there for 21, he asked Dr. Kapur to provide him guidance in general and, specifically, to tell him whether a way had been found to institute microeconomic reforms in millions of enterprises.

Dr. Kapur stated his full agreement with Mr. Kalotra's assessment that 100 companies constituted a modest initial step, but he suggested that leadership is the most important element when embarking upon change. IT had, after all, started off in India with only two or three companies. CII believed that creating 100 leader companies, which might in turn have hundreds of suppliers, would get

---

[6]National Academy of Engineering, *The Engineer of 2020: Visions of Engineering in the New Century*, Washington D.C.: National Academy Press, 2004, p. 4.

change under way: that if a few hundred companies were created, thousands more might come about. However, "it would be very pompous of me to say that we're going to have thousands of companies involved in this change movement overnight," he remarked, adding: "It's not going to happen." Just changing people's mindset is difficult, and CII, being cognizant of the Indian context, needs to take a very realistic view.

In response to Mr. Kalotra's question about clustering, Dr. Kapur said that the Indian government has begun sponsoring programs in which rival firms collaborate on precompetitive research. New technologies developed by the companies—with the participation of Indian research institutions and, in some cases, U.S. academics or other outside resources—would result in common intellectual property rights in a generic area. The firms could then find applications on their own and compete at the product level. India's Core Group of Automotive Research, or CAR, was one such collaborative program. Dr. Kapur added that the clustering program he had referred to is aimed at helping noncompeting companies learn the processes of change rather than at specific industrial applications.

He then turned to Dr. Sinha's reference to the NAE report regarding "customerization." Changeover times are extremely important for going from mass customization to one-off production without loss of efficiency. Indian firms are learning, through programs already under way, manufacturing-cycle efficiency needed for success in handling "design of one" or "production of one."

Turning to Dr. Wessner's question about SMEs, Dr. Kapur stated that the real challenge was motivating SMEs to do R&D. CII is finding that as companies' internal processes and managements improve, they begin to move in the direction of doing R&D and being involved in innovation. It was here that the clustering approach comes in. One illustration is CII's Chennai cluster of six companies in the leather industry, which—undoubtedly to the surprise of many—is a larger exporter than the auto components sector. Dr. Kapur termed the benefits beginning to accrue to these six companies, which had agreed to work together in the cluster even though they were competitors, to be "just amazing." As he had said in conversation with Dr. Patel earlier, where help from the World Bank and similar institutions was really needed was in funding programs for SMEs; establishing such programs would represent a major step forward.

Before taking a final question, Dr. Patel wished to note that the question Mr. Kalotra had raised regarding the role and timing of macroeconomic, microeconomic, and enterprise reforms was not susceptible of a clear-cut answer. Even if a discussion could be focused on one specific subject, such as the day's theme of innovation, it was always the reality on the ground that many things had to take place in tandem and at the same time. Deputy Chairman Alhuwalia, who had dealt with this issue from 1991 onward, might have had a better answer, allowed Dr. Patel. His own view, however, was that India's macroeconomic reform program had to continue and that it had to take into account all the issues regarding the overall economic environment in India so far raised at the symposium.

He then listed a number of binding constraints that had been mentioned:

- the low level of education, which has implications for what macro policy needs to do to create the kind of skilled labor and skilled human resources that India needs to grow at 10 percent per year;
- the challenge of achieving results in the context of the "two Indias," a phenomenon that has macroeconomic implications because, if it were not addressed, the people or regions or states left behind would become a source of instability.
- the lack of infrastructure, which refers not simply to building more roads but to creating a macroeconomic framework that would allow investment from the private sector to fill the strategic infrastructural gaps that existed.

The answer to the question, therefore, was that owing to the complexity of actual conditions, India's macroeconomic managers had to continuously create the kind of space and environment that would allow the scale-up of the programs that had been described. He then asked for one more question.

Sujai Shivakumar of the STEP Board, referring to Dr. Dahlman's statement that the level of private-sector R&D in India was relatively low, asked whether Dr. Kapur had an explanation for this and whether anything was being done about it. In addition, he asked Mr. Subramanian whether there were lessons to be learned from those among the Indian states that appeared to be doing better than the others.

Mr. Subramanian acknowledged that some states were indeed doing better than others, ascribing this in part to the history of the Mogul and British periods and specifically mentioning the contrasting development of the coastal and interior regions during the latter. Additionally, many of India's southern states had far better systems of basic education and nutrition, a fact related to these states' response to public developmental issues under democracy. But this was also linked to broad macroeconomic development and came back to fundamentals: Without concentration on public health and primary education, nothing else will flow. Playing a role as well was the fact that, while some states are still absorbed with issues of caste and other local concerns, others have moved into developmental issues on a slightly higher plane. Dr. Shivakumar's question was an important one, he said, because most of the themes affecting the average citizen are within the authority of the states, with the central government doing only so much to direct traffic in those areas.

Dr. Kapur, in response to Dr. Shivakumar's inquiry about research spending by the private sector, noted that R&D is a recent phenomenon in India. Before reform took hold in 1991, restrictions on investment had been significant. Even today, most manufacturers there are involved in one form or another of contract manufacturing and thus have no real need to do R&D. In view of this, current R&D spending could be considered fairly good. Secondly, a company such as

Sona, which is in the metal-forming business, does all its design, analytical work, and simulation in virtual space, where costs were very low. He therefore expressed doubt that India's private sector needs to spend a much higher percentage of its revenue on R&D.

However, the real issue, he stressed, is that R&D is required by companies creating their own products and processes, their own intellectual property, rather than by contract manufacturers. He noted that Swati Piramal was to speak later in the day about the R&D efforts that Nicholas Piramal India Limited is making because such efforts are now required of pharmaceutical companies whereas they had not been in the past. A change was taking place, everyone had begun doing R&D, and this would be reflected in spending growth.

Dr. Kapur then turned to the issue of skill development, which, he predicted, would impel great movement in public–private partnerships. India possesses a phenomenal physical infrastructure in its Industrial Training Institutes (ITIs), but owing to a lack of state funding they have unfortunately been allowed to decay over the years. A shift is under way, however: The previous year, India's Finance Minister challenged industry to create partnerships with ITIs, and many firms are beginning to work with their local institutes. His own state, Haryana, has relinquished exclusive ownership of its ITIs, giving industry a much larger role in them. The private sector will assume increasing responsibility for skill development because, as Dr. Kapur put it, "this is our requirement and we need to do it for ourselves."

Dr. Dahlman said that if India could leverage the resources it had within its own national economy, it would make tremendous progress. Critical to this was Indo–U.S. collaboration, which has the potential to become "a very nice, symbiotic strategic alliance." This prospect, he stated, lent excitement to the day's meeting, and it would, he hoped, provide the impetus for nailing down specific opportunities for partnership.

Dr. Patel concluded by thanking the presenters and questioners alike for a very interesting session.

India's Changing Innovation System

Introducing India's minister of science and technology, Kapil Sibal, and John Marburger, director of the White House Office of Science and Technology Policy, Dr. Wessner expressed, on behalf of the National Academies, not only delight but also gratitude to the minister for making the trek from New Delhi to Washington and to the president's science advisor for forsaking his heavy schedule to take part. He added that the scope and intensity of the U.S.–India relationship, so frequently mentioned during the morning's discussion, was reflected in their joint presence on the dais.

INTRODUCTION

John Marburger
White House Office of Science & Technology Policy

Dr. Marburger, in his remarks introducing the minister, observed that Americans have learned that it is important to listen to India because important things are happening there. All are aware not only that President Bush's visit to India earlier in 2006 marked a turning point in relations between the countries, but also that it has resulted in historic and widely publicized agreements. The present symposium, he said, is one of many activities that have taken place or been planned since then to realize the joint vision that President Bush and Prime Minister Singh put forward on that occasion. The nations' long partnership has entered a truly remarkable time.

Their vision, spelled out in the communiqué that capped the meeting of the two heads of state, encompasses five themes that are strongly linked to and deeply dependent upon the technical capabilities of their countries: economic prosperity and trade, energy security and a clean environment, innovation and the knowledge economy, global safety and security, and deepening democracy and meeting international challenges. While these themes could be said to summarize the challenges and aspirations of all nations participating in the globalized economy of the 21st century, they hold a special significance in the case of each. The nature of the U.S. relationship with every country depends on the unique characteristics of that country, its capabilities, its position in time and space, and the challenges it faces—factors that also shaped the U.S. response to the partnership. It was this uniqueness, Dr. Marburger explained, that is behind the necessity of his hearing what Minister Sibal had to say.

Scholars speak of two distinct ways of understanding human affairs: the "diachronic" or historical approach, which traces the origins of a situation, and the "synchronic" or snapshot approach, which seeks the structure inherent in a pattern of events at a given moment. During his remarks at dinner the previous evening, Minister Sibal had suggested that the historical approach did not suffice to describe or explain what was currently happening in India, particularly in regard to its relationships with other countries. He had urged his audience to look instead at the geography of international developments, as well as at the distribution of economic activity and its technical basis in space rather than in time. Change was occurring too rapidly, he had inferred, for guidance based on history alone to be reliable.

Dr. Marburger strongly identified with this point of view. Americans would not be able to see the course of their future relationship with India clearly by examining the trajectory of past interactions. The present differed too radically from anything known before, and direct, real-time interaction among parties was required. Just such an opportunity was being provided by the day's symposium, and he thanked his Indian colleagues for taking the time to bring word of the extraordinary developments afoot in their part of a new global economic geography that they had done so much to create.

The symposium was also providing an opportunity, at an important moment in time, to get a synchronic snapshot and to ponder the patterns that it revealed. Scientists tend to equate innovation with new ways of looking at their fields or with new tools—new instrumentation—for broadening the opportunity for discovery. In business, innovation more often means introducing new ways to solve a problem, satisfy the needs of a market, or deliver a product more efficiently. Explaining the difference, Dr. Marburger said that scientists try to map out the structure and properties of nature, whose laws are relatively constant. "It's not that nature stands still for us," he said, "but at least it doesn't change its face, particularly from day to day." Once made, therefore, innovations in science can last for a long time. In contrast, rather than mapping out an unchanging nature, economic activity involves grappling with a continually changing social reality whose varying circumstances require constant attention. Innovations are not "once and for all," but transient, serving for a brief time and then losing their potency or their market.

For this reason among others, according to Dr. Marburger, innovation was not a zero-sum game, and the United States had no need to fear that it would lose anything by working with other countries to develop their innovative capacity. The geography of the global economy was such that different innovations were required for the unique conditions of each separate region—something that, while true even within a given country, was especially true among countries. "We should be particularly eager to work with India, which is the world's largest democracy and increasingly important to our own innovation economy, to magnify our mutual capacity to address our respective problems," he said.

Focusing specifically on the man he was introducing, Dr. Marburger said he had been struck, during their brief interactions of the previous year, by the combination of talents that Kapil Sibal had brought to the position of minister of science and technology. "At every international conference I go to, I see his face and hear his booming voice, and I usually hear something that I didn't expect to hear—some insight that greatly impressed me," he said, adding: "It doesn't take too many Sibals to make a dynamic country."

First elected to Parliament in 1998, Kapil Sibal served as the official spokesperson of his party, the Indian National Congress, during the 1999 and 2004 parliamentary elections. A former cochairman of the Indo–U.S. Parliamentary Forum, he assumed the post of minister of science, technology, and ocean development in January 2006. "Candid and forthright in his political views, Sibal has often publicly criticized the Congress Party for some of its policies," said Dr. Marburger, quoting material provided by the Indian Embassy. The president's science advisor then offered a comment of his own: "That's brave. It's impressive. It tells us something about India, and it gives us pause in this country regarding our own systems."

Born in 1948, Minister Sibal is well known in India for pleading cases before its Supreme Court. He came to the limelight in 1993, when in the capacity of attorney he addressed Parliament's Lower House, the Lok Sabha, for three consecutive days during the historic impeachment of a sitting justice of the Supreme Court, the first such proceeding against a member of India's superior judiciary. He holds a Master of Arts degree from St. Stephen's College of Delhi University and a Master of Laws degree from Harvard Law School. He joined the bar in 1972, has served as additional solicitor general of India, and was thrice elected president of India's Supreme Court Bar Association. In 1991, he led the Indian delegation to the United Nations Commission on Human Rights (UNHCR) in Geneva, and he has been a member of the UNHCR Working Group on Arbitrary Detention. "It's very impressive that India has chosen a prominent attorney who has an interest in human rights to be its science minister," Dr. Marburger reflected, adding: "Not that I don't have an interest in human rights myself, but it's an unusual thing and it speaks to the unique qualities that Minister Sibal brings to his work." Currently, he was serving on the Governing Body of St. Stephen's College and on the Board of Management of the Indira Gandhi National Open University, both located in New Delhi.

Ceding the floor to Minister Sibal, Dr. Marburger said that he would look forward to hearing the minister's words of wisdom on what he hoped would be his frequent visits to the United States, just as all present were looking forward to hearing him speak at that moment.

INDIA'S CHANGING INNOVATION SYSTEM

Kapil Sibal
Ministry of Science and Technology

Minister Sibal began his remarks by thanking Dr. Marburger for the introduction and promising not to let him down by failing to say something that he had never heard before. He explained that his preference, expressed the previous evening, for the geography of science over the history of science stemmed from the conviction that we at the dawn of the 21st century are no more able to imagine the changes that will take place in the coming hundred years than observers of a century ago were able to envision life as it is now.

The challenges of the past century were far different from the challenges that our civilizations are to face in the one just begun. The 20th century was a century of conflict: Empires created in the 19th century were on their way to being dismantled before its close, but it was in the 20th century that the dismantlement had been completed. The driving force in both building and dismantling empires, Minister Sibal declared, had been technology. In the march of civilization through the 20th century, force and velocity were at the heart of technological development. "Force and velocity became, in a sense, policy determinants," he said. "Force and velocity were the reason for change."

In the 21st century, however, all would be different. "Force will have no role to play; markets will." In an ever-expanding global economy, the challenges would relate to the availability of water or energy, to the environment, to disease, to hunger, to poverty, and to natural disasters, with many others on the list as well.

Referring to the title of his talk, the minister posited that a prerequisite to understanding India's changing innovation system was an understanding of America's changing investment needs, as the two went hand in hand. Why were Western countries, among them the United States, looking for markets? Because today's consumer wants the highest quality product at the lowest possible price. Meanwhile, the value of physical assets owned by multinationals have been on the decline over the previous 20–25 years, but the value of their nontangible assets—their intellectual property—has been increasing. These facts are at the heart of contemporary technological development, which is driving economic growth everywhere in the world.

At the same time that the world's multinationals are looking for access to larger markets, the world trade regime (represented by the World Trade Organization) allows countries to lower their tax and tariff barriers in order to provide that access. Countries such as India and China have, therefore, been able to play a very significant role in the changing investment scenario. The challenge before them was to collaborate with the West—this is imperative—while at the same time creating intellectual property of their own through innovation. "And innovation is not about technology alone," Minister Sibal stated. "Innovation is about the use of ideas and the use of knowledge in its application for change. And when that happens, then it's a win-win."

Minister Sibal noted that conference participants could at the moment be feeling either "euphoric" or "exceptionally pessimistic" about India, since they might have drawn either of two opposite conclusions from the morning's presentations: that the country was on the fast track or that it faced formidable obstacles to progress. However, this could be the case regarding any of the world's nations, depending on one's point of view. With a middle class that, at 300 million, might be larger than that of any other country, India has not done too badly.

Nor should one forget that the United State had more than 200 years of history behind it when considering where technology has taken that nation, the institutions it has built, or the systems it has put into place. America's history was not to be compared with that of India, where liberalization started in 1991 and which has had just 15 years to master developmental processes and reach levels of excellence that countries such as the United States have taken 200 years to build up.

Alluding to concerns about pollution levels in Delhi, Bangalore, and other parts of India and about infrastructure development, Minister Sibal declared: "Nothing can happen overnight." He recalled hearing as a young student about the levels of pollution in Chicago, New Jersey, and many other places in the United States, and he remarked that the country had dealt with such problems through building institutions, through innovation, and through applying technologies for change. "The expectation that this must happen in India tomorrow," he protested, "is a little unfair"—even if one wished, as did he, that "it had happened yesterday."

India, perched on the cusp of great opportunities, wished to collaborate with the United States so that the country could show Americans all it had to offer them while at the same time meeting its own challenges. The minister equated solving India's problems to solving the problems of the world, adding: "That's what India is all about." The country has a six-lane highway over which traffic moves with great speed, but it also has bumpy, single-lane roads. Its challenge was to turn the latter into the former.

Expressing pleasure at the prospect, discussed earlier in the day, of India's receiving help with environmental and energy needs through partnering with the United States in such programs as FutureGen, the minister reiterated India's desire to help U.S. businesses with access to markets. "India wants to march along in the comity of nations as an equal partner, and India would like to collaborate with countries as an equal partner," he stated. "This is verily the win-win situation that we must talk about."

To illustrate the exciting changes currently taking place in India, Minister Sibal turned to health care. Unlike in the United States, 80 percent of India's health sector is controlled by private industry. Visitors to India are finding some of the finest hospitals and health care, which are available to all those able to pay, and a great deal of health tourism is taking place: Foreigners come for care because India offers lower costs and high-quality health care in superb hospitals. At the same time, entire segments of the Indian population have no access at all to health care. It was not advanced technology that can solve the problem of access to health care in India, according to the minister; instead, solutions are needed that, while technologically based, are accessible and affordable.

A partnership recently undertaken with a private urban hospital delineates the kind of change the Planning Commission is seeking. Minister Sibal has proposed that the hospital care for a cluster of villages and suggested that it set up a "medical kiosk" to serve their population, since poor people could not be expected to travel to Delhi for treatment at its facilities. "'Let the rural folk come to that little medical kiosk,'" he recalled telling hospital officials, "'and, through remote satellite technologies'—which we have in India—'let them be diagnosed by people sitting in the urban centers.'" The Ministry of Technology is covering the necessary investment in medical hardware; the hospital is making available the doctors. Although technology-based, this solution to the problem of medical care in rural areas was "not high-tech," and it could be implemented at low cost.

Similarly, it was thanks to what Minister Sibal called "very simple technology" that clean water is being made available to some populations for the first time. Many of the 400 million who live along India's coasts have no reliable access to drinking water. He was soon to inaugurate a thermal desalination plant in Chennai whose technology would exploit the difference between the temperatures of water at the ocean's surface and at a depth of 200 meters by using surface water for flash evaporation and deep water for condensation.

The minister's message for the rest of the world, and specifically for the people of the United States, was that collaboration is required at both ends of the technology spectrum. India needs to collaborate with the United States in bringing high-level technologies to bear on the larger problems of the world: in the context of energy, for example, atomic power generation, photovoltaics, hydrogen cells, or next-generation, zero-based coal technologies. However, India also is in need of very simple technologies to improve the lives of 600 million of its people, 500 million of whom are living on less than $2 per day. Such technologies are crucial because, as he put it, "the object of technology development is ultimately economic growth and raising the living standard of all, not just a few."

The world at large also has a stake in the success of these endeavors, stressed Minister Sibal, and it is related to the United States' changing investment needs. About 90 percent of India's economy is already free, and the prime minister has promised publicly that tariffs would continue to be lowered until they reach levels of the Association of Southeast Asian Nations, which would improve access. India represents a huge market, and if it is able to improve the lot of the common man and to put more money into the hands of India's rural folk, they would have a larger capacity to buy consumer goods.

At present, 60 percent of India's population—more than 600 million people—live in rural areas. India leads the world in the production of milk, sugar, and tea; it is the world's second-largest producer of wheat and rice, and the second-largest agricultural producer overall. Yet India's per capita per acre productivity lags far behind those of the leaders. Of Indians making their living from agriculture, 60 percent are marginal farmers without the wherewithal to invest in the land. A large part of the agricultural sector lacks any access to technology. Incumbent upon the country, therefore, is to improve its best practices. This includes ensuring that farmers get good seeds, and that they apply multicropping patterns. Farmers also need clear and reliable market access, and this calls for getting rid of commission agents, and making available a cold chain from the field to the market. By using best practices and simple technologies to raise productivity levels, India could become by far the world's largest producer, and a major exporter, of agricultural commodities. And if it invests in agro-based industries, a conspicuous strength of the United States, it would be able to sell value-added products to the rest of the world and bring about economic growth.

With some of India's economic sectors already doing well, Minister Sibal noted, questions have arisen as to why employment opportunities remain inadequate. "Employment opportunities do not come by education alone," was his answer. "Employment opportunities come through growth." Why are there employment opportunities in the IT sector? Because growth in the IT sector, which has evolved into a $20 billion to $30 billion industry, has spurred enrollment in IT programs at the country's educational institutions. Why do employment opportunities exist in the biotechnology sector? Because that sector's annual turnover has reached $1.5 billion. Why is there growth in the pharmaceutical sector? Because it has become a $20 billion industry.

Minister Sibal expressed his hope that India's pharmaceutical companies, already important producers of generic and bulk drugs, would capture 90 percent of the U.S. market for the latter in the years ahead. However, India's pharmaceutical industry is doing much more than exporting bulk drugs. It is innovating, producing its own products and lead molecules; it is acquiring patents; and it is acquiring companies all over the world, a subject that Swati Piramal of Nicholas Piramal India Limited would address later in the program.

While acknowledging that, as some had remarked, Indian levels of investment in R&D are low, he stated that they would always be low unless there is economic growth. Why was it that the level of R&D investment in India's pharmaceutical sector has grown to almost 6 percent and that its 20 top companies are investing more than 6 percent, or that some of its companies were investing as much as 20–24 percent of sales in R&D? It is because that sector is doing well. Similarly, why is India's automobile component industry doing well? Because there is economic growth. "Investments come through economic growth," the minister stated. "Once there is economic growth, there will be more investment in R&D."

Emphasizing that India has to continue with its process of economic liberalization in order to provide opportunities for investment, Minister Sibal asserted that a partnership between India and the United States is "vital" for the economic growth of both nations, whose combined population was greater than that of China. "Just imagine the market that we can have access to!" he said. "And not just the market of people in India and the United States of America, the world market." Putting forward areas for collaboration, he noted that India is a leader in space technology and satellite communications systems, having its own capabilities in missile technology. A U.S.–Indian partnership in those areas would ensure world leadership. India has also become a leader in vaccine development, providing 90 percent of the world's measles vaccine and selling hepatitis B vaccine, which at one time cost more than 900 rupees per dose to import, at only 16 rupees per dose.

For Americans, the challenge is to recognize that India as a nation faces challenges and that the United States as a nation can realize opportunities from those challenges. "That is a partnership that matters," the minister stated, urging: "Let's meet the challenges. Let's create the opportunities." Attributing to Napoleon the statement "Science is the god of war," he offered an alternative: "Let science be the god of peace and of prosperity."

DISCUSSION

At the conclusion of Minister Sibal's speech, Dr. Marburger suggested that they entertain questions from the audience, a proposal to which the minister agreed.

Anita Goel of Nanobiosym said that while both speakers had been quite eloquent regarding the need to work together on a model for innovation, it was not clear what specific roadmap would enable India and the United States actually to join hands in building it. What role, she asked, did each see for government, for academia, and for private firms? What would be the interplay among them? Did the speakers envision something like a self-assembly model, in which the government would create only an environment and the system would self-organize? Or did they see government taking a more proactive role in creating these collaborations?

Dr. Marburger answered that the first step had been taken by Prime Minister Singh and President Bush in focusing attention on not only the need but also the opportunities for partnership between their two countries, and in raising its level of priority within government. Just the previous fall, he had joined Minister Sibal and Secretary of State Rice in signing a Science and Technology Agreement that set out a framework identifying broad areas for collaboration. The next step would be to identify agencies whose programs fit into that framework and specific mechanisms that would allow the governments to invest in partnership programs that could bring value to both sides.

While government expenditures by themselves would not have the greatest impact, they could lower barriers and pave the way for other, larger ventures. With economic activity already at a very high level, the questions have become which components afford the highest leverage and what obstacles remain to even further development in the future.

Minister Sibal added that, in his view, effort is needed at three levels. Number one is at the government-to-government level. Here, the two countries already have in place such great enterprises as the Bi-national Science and Technology Endowment Fund, Initiative on Agriculture, and High-Technology Cooperation Group. However, it is necessary to move beyond that to a second level of collaboration, and that this is in progress as well. The minister said that he has visited a large number of U.S. universities, and each desired to arrange to do collaborative work in niche areas of expertise with institutions in India's university system, among its autonomous research institutes, or both. Such agreements already exist: The University of California had joined with Indian research institutes in an undertaking requiring a $10 million annual investment.

The third level at which India needs to move forward is to make sure such investment and collaboration are taking place in a friendly environment so that more people and institutions would be attracted to India. A product patent regime has already been put into place for the pharmaceutical sector. In the Parliament's winter session, legislation along the lines of the American Bayh–Dole Act would be introduced with the goal of guaranteeing to the scientific community working in Indian educational institutions ownership of intellectual property that it created. In addition, India is establishing special economic zones where huge excise and tax benefits are to be given and is also setting up biotechnology parks.

A biotech development strategy being implemented allows government funding of start-up companies. "We are going to public–private partnerships in a big way, giving money to small and medium-scale enterprises in the biotech and other sectors to make sure they do new kinds of research for new products," the minister stated, inviting the audience to "come to India and see what's happening."

Som Karamchetty of SomeTechnologies introduced himself as a product of globalization: Originally an Indian citizen, he subsequently became an Australian citizen and was now an American citizen. He recommended that India "translate" technology it might see or acquire from abroad rather than applying it without thought to its appropriateness or simply copying it. "Please do not develop roads like those in America," he urged, characterizing the Capital Beltway ringing Washington, D.C., as a six-lane "parking lot." He pointed to contrasts in telecommunications infrastructure and distribution of drinking water that also differed in the U.S. and Indian contexts and might affect development strategies. In light of such differences, he asked, who would distinguish between technologies that could simply be taken over and those that might need to be modified or substituted, and how would this be done?

Dr. Marburger said that the needs of the United States and the needs of India, while they might be distinct, did in some respects overlap and reinforce each other. India could help the United States to add value to its products and, by doing so, enrich itself so that it was able to invest in upgrading its own infrastructure and broadening its markets.

Minister Sibal, concurring with Dr. Marburger, stated that collaborative activity in India has a twofold structure. At one level, the 300 or so Fortune 500 companies investing in R&D in India are using Indian talent—80,000 engineers and scientists—for the creation of intellectual property targeted at their home markets. At the other level, investments in the United States are underwriting the quest for the kinds of affordable, accessible technology solutions required in India. Both would, and should, continue to happen.

Hiten Ghosh of Hughes Network Systems introduced himself as a representative of the 40,000 IIT alumni who had become U.S. citizens. Just as there were opportunities for the United States in India, he said, the two nations have many common problems, so that solutions that might be arrived at through collaboration could benefit both. The application of technological best practices that had helped fishermen in India, for example, might also be of benefit in the Rust Belt and in some U.S. rural areas. While there might be a difference of scale, some elements of the retraining problem that both countries face are constant. Addressing both speakers, he asked how connectivity and collaboration could be used to address common problems.

Minister Sibal said that the areas of collaboration on global issues that he set out for two countries—water, energy, environment, disease, and natural disasters—would have ramifications worldwide. India and the United States both have the human resources and the technology to deal with those issues even if their skill sets are different. The United States might work on hydrogen cells, India on photovoltaics, he said, picking an example from the "multifarious" energy sector but stressing that it could be extended to all others. Under collaboration of the sort needed, the two countries could pool their resources but work in different areas in order to provide solutions that could be applied anywhere in the global marketplace.

Dr. Marburger cited as common issues energy, environment, some areas of public health, and water management, although he added that the dimensions might be slightly different in each of the countries. The prodigious intellectual talent that India has sent to the United States is helping in the quest for solutions to these problems. The education being brought from the IITs is value added to the U.S. economy and culture, and it is highly valued in turn.

Dr. Telang of Howard University noted that Minister Sibal's presentation sent a very powerful message and at the same time issued a challenge. The quote from Napoleon with which the minster had concluded his keynote address made clear the potential of science and enterprise. Alluding to the presentation of Mr. Subramanian in the previous panel, who had declared that much of the policy and implementation responsibilities relating to innovation lie with India's state governments, he inquired if there is sufficient dialogue between the central government and the state governments.

"There is no dialogue between me and Mr. Subramanian, I can confess to that," the minister quipped. Then, saying he considered the issue a federal one, he added that India's central government is in fact collaborating with the states through contact with state chief ministers and through conferences—and even if the U.S. federal structure is far stronger than India's, Indian states do not disregard federal objectives. While acknowledging that the federal government might not have laid sufficient emphasis on some of the challenges that Mr. Subramanian talked about, Minister Sibal contended that the picture was not as dark was made out to be.

Bringing this session to a close, Dr. Wessner expressed his appreciation to Dr. Marburger and Minister Sibal. Calling the minister's list of the new initiatives "impressive," he reminded the audience that the conference represents an opportunity for mutual learning. Both countries, he noted, face new global challenges and see the need for policy change as well as collaboration to adapt to this new competitive environment.

Growing the Science and Technology, Research, and Innovation Infrastructure

Moderator:
George Atkinson
Department of State

Dr. Wessner introduced the panel's moderator, George Atkinson, science adviser to U.S. Secretary of State Condoleezza Rice, as a friend of the National Academies, a friend of India, and a friend of science.

Dr. Atkinson welcomed the audience, fresh from hearing the inspiring words of Minister Sibal and Dr. Marburger, to a new chance to learn more about the specifics that were to engender the very ambitious program being undertaken by the United States and India. The present panel could be expected to focus on two issues—cooperation, and mutual benefit and listening—that had been raised by the previous speakers. Recalling Minister Sibal's reference to the geographic character of science, Dr. Atkinson opined that the transitory nature of leadership in science was evident to all. Europe had been at the forefront of science and technology in the 19th century and the early part of the 20th century, the United States had taken the leadership position in the recent past, and many now agreed that the future belonged to those able to find modes of collaboration.

R. A. Mashelkar, the first of the panel's three speakers, is the director general of India's Council on Scientific and Industrial Research (CSIR), the largest chain of publicly funded industrial research and development (R&D) institutions in the world. He is also president of the Indian National Science Academy and, since 2005, a foreign associate of the U.S. National Academy of Sciences.

RENEWING THE NATIONAL LABORATORIES

R. A. Mashelkar
Council on Scientific and Industrial Research

Dr. Mashelkar began by recalling the evening of April 22, 2006, when, on the same stage, he signed the register signifying his membership in the National Academy of Sciences. The Academy's president, shaking his hand, said that the Indian flag would be displayed in his honor. This was "an unbelievable matter of pride," said Dr. Mashelkar, who added that his daughter had taken more photos of him with the flag that night than signing the register. And little did he realize then that less than two months later the Indian flag would again be flying in the auditorium and he would have an opportunity to speak from its podium. He expressed his thanks.

Dr. Mashelkar proposed to tell the story of the Council on Scientific and Industrial Research within the framework of a generic discussion of national laboratories that would also include a more specific issue, that of their renewal. His opening point was that "context decides the content," and that the context changes not only from country to country but, in a given country, with the passage of time. For this reason, laboratories designed and established to serve a particular national purpose necessarily change. As an example, he cited the Global Research Alliance, which comprises chains of national laboratories: CSIR India, CSIR South Africa, CSIRO Australia, Fraunhofer-Gesellschaft Germany, VTT Finland, DTI Denmark, TNO Netherlands, Battelle U.S., and SIRIM Malaysia. In line with the diversity of their home countries, these labs serve different purposes, and they had changed over time as well.

Personal experience, acquired not only in his own country but in some others as well, would shape Dr. Mashelkar's remarks. Over the three decades that he had spent with India's national laboratories, he had helped restructure industrial R&D institutions in South Africa, Croatia, Turkey, Indonesia, and China. He chaired the committee that reviewed CSIR South Africa in the period 1997–2002 and, together with a member of the audience, Vinod Goel of the World Bank, worked on projects in Croatia between 2002 and 2004 and in Turkey between 2001 and 2005.

What are national labs supposed to deliver? Dr. Mashelkar's answer was private goods and services, public goods and services, strategic goods and services, and social goods and services. He summarized the activities that take place under these four categories as follows:

• **Private:** creating new intellectual property, licensing and commercializing technologies, making a country's industry globally competitive.

• **Public:** generation and dissemination of scientific research, creation of scientific and technological manpower.

• **Strategic:** finding technological solutions for national security, strategic positioning of industry, and representation in global affairs. As an illustration, he noted that research by CSIR of India has produced, for use in the wings of the country's light combat aircraft, carbon-carbon composite technologies that "are not available to [India] for love or money." Radiation-shielding glasses for atomic energy work and microwave tubes for India's space program are other strategic goods designed and fabricated by CSIR.

• **Social:** providing for employment, health care, drinking water, and other fundamental needs of those below the poverty line.

Dr. Mashelkar proposed to take the audience rapidly through a study of the transformation of CSIR in India. Upon taking its helm in 1995, he was told that he would face different expectations from each of a number of stakeholders: the average citizen, women and children, farmers, industrialists, academics, politicians, bureaucrats, and the military. After a decade, how did four such groups—business, management experts, leading scientists, and the political leadership—view the direction in which CSIR had gone?

• **Business** appears to be content, as evidenced by a cover story in a leading publication, *Business India*, that said: "CSIR labs have been transformed by the power of enterprise and proactive management. . . . R&D is, at the end of the day, a commercial activity. The CSIR labs are going places with this idea to inspire them."

• **Management experts** have examined how organizations have transformed themselves during the first decade of India's economic liberalization, which began in 1991. In a book edited by the late management guru Sumantra Ghoshal, a chapter on best practices in managing radical change mentioned, in addition to a variety of prominent firms, only one public institution: CSIR.

• **Leading scientists.** In his 2003 book *The Scientific Edge*, the famous Indian astrophysicist Jayant Narlikar placed CSIR's transformation among India's top 10 achievements of the 20th century, in a league with the Green Revolution and the work of the country's Atomic Energy Commission on nuclear power.

• **Political leadership.** Indian Prime Minister Manmohan Singh, the ex-officio president of the CSIR Society, said in 2004: "I would like to congratulate CSIR for the remarkable transformation into a performance-driven and user-focused organization. I am happy to see that CSIR is flying higher and further." India's previous prime minster, Atal Bihari Vajpayee, had said in 1997: "CSIR has regained its dynamism and prestige, besides showing itself to be capable of standing up to the challenges of liberalization and globalization."

Equally important and a particular source of pride, Dr. Mashelkar stated, was that CSIR was being used as a model of institutional transformation by the World Bank. Alfred Watkins, the head of the Bank's Europe and Central Asia Region, has written that in the course of his work, he "recommended CSIR as a model of how countries can harness their top-quality scientific research institutions to the task of industrial technology development, innovation, and global competitiveness."

But what is the mission of CSIR? Dr. Mashelkar read its most recent formulation: "To provide scientific industrial R&D that maximizes the economic, environmental, and societal benefits for the people of India." The domains mentioned—economic, environmental, and societal—define a "triple bottom line" whose management "through high science and technology," he stressed, constituted "The Big Challenge" for CSIR or, in fact, any set of national laboratories.

An illustration of how this challenge was being met in India could be found in the development of a method of silver sulfadiazine (SSD) microencapsulation on collagen-based biomaterials by its Central Leather Research Institute (CLRI). While this was ostensibly a "mundane research institution," he said, it specialized in the science of collagen and had come up with remarkable systems for healing burn injuries. Noting that national labs are often judged harshly when the criteria used are revenues, patents, or publications, Dr. Mashelkar projected a photo of young children whose burns had been treated with CLRI's collagen dressings and suggested attributing value instead based on the smiles covering their faces. "We need to recognize, particularly in a country like India," he declared, "this larger context in which the national laboratories operate."

Thanks to its efforts to maintain the quality of its science while in pursuit of its mission, CSIR had also done quite well when it came to publications and patents. The number of basic-science papers that its researchers had published in Science Citation Index (SCI) journals had climbed rapidly, from 1,700 in 2001 to 3,018 in 2005. Currently, one in six Indian papers that appeared internationally in peer-reviewed journals comes out of CSIR, which leaves its production just 10 percent short of that of the Indian Ministry of Science, the country's leader. At the same time, the number of U.S. patents granted to CSIR has soared, from single digits through most of the 1990s to 145 in 2002 and close to 200 in 2005. And among Patent Cooperation Treaty applicants from developing countries, in 2002 CSIR tied for first place, at 184, with South Korea's Samsung Electronics—the latter's R&D budget being, he said, 10 times larger. CSIR scientists had, in fact, been recognized in the media for several breakthroughs in 2005.

It was a polycarbonate patent developed in one of CSIR's laboratories that led to the founding of GE's Jack Welch R&D Center in the state of Karnataka. A partnership had begun when the company licensed the patent a dozen or so years before, and then Welch asked one day: "If they're so good, why aren't we there?" Any foreign company interested in establishing an R&D center in India now wants to visit the Jack Welch R&D Center, which had become a sort of showplace but is just one of many that U.S. firms maintained in India and that had led to the country's emergence as a global R&D hub.

Recalling a challenge articulated earlier by Minister Sibal—delivery of high-quality products at low prices—Dr. Mashelkar noted that meeting it would put products within the reach of a larger part of the population. This could be achieved only through very inventive use of high science and technology. As an example of an affordable solution, he cited the application of a "process for the preparation of ultrafiltration membranes of polyacrylonitrile, using malic acid as an additive," the object of a 2005 U.S. patent based on precipitation volumiza-tion, a unique technique allowing the creation of 20-nanometer pores that had been developed at India's National Chemical Laboratory (NCL). It had resulted in devices able to filter not only bacteria but viruses from drinking water at a cost of one-tenth of a cent per liter, and the more than 2,000 deployed included some in rural villages without electricity that were operated by hand pumps. "This is where India and Bharat, or Hindustan as we call it, coexist in some sense or other," he said, alluding to the image of the two Indias invoked by previous speakers. Such things have become possible thanks to a new vision and strategy embodied in a white paper entitled *CSIR 2001*.

Dr. Mashelkar then outlined cultural differences separating the CSIR insti-tutes from industry:

- **time horizon**—long term for institutes, short term for industry;
- **financial structure**—institutes based on cost centers, industry on profit centers;
- **products**—institutes generated packages containing knowledge and in-formation, industry is interested in salable goods and services;
- **basic orientation**—scientific novelty for institutes, market attractiveness for industry; and
- **focus**—institutes looked at perceived needs, industry at market needs.

These differences represent a significant challenge for industrial R&D labo-ratories to meet.

With the help of the World Bank, CSIR has made attempts to bridge these gaps. As part of an industrial technology development program whose first phase started in 1991, marketing teams were created in each laboratory, decision mak-ing was devolved, specialized business development consultants were brought in, members of the CSIR staff were allowed to serve on boards of directors in the private sector, awards were given for marketing and business development, and knowledge was used as equity. At the same time, CSIR created financial incen-tives, linking performance to budget allocations, offering incentives to scientists, and establishing laboratory reserve funds. Under the last, labs were permitted to retain and carry forward surpluses based on a percentage of earnings, which they were then to use for development purposes as they saw fit. The amount of earnings accumulated in these funds had risen along a straight line that had begun just above zero in 1992–1993 and reached 4.5 billion rupees by 2005–2006.

To illustrate a final aspect of this cultural change, the move away from reverse engineering and toward what Dr. Mashelkar called "forward engineering," he offered the history of work in catalysts done at NCL. While that lab was commercializing a "me-too" catalyst, dimethylaniline, as late as 1978, by 1986 it had developed catalysts on xylene isomerization that, he said, were superior to Mobil's catalysts. Reverse technology transfer to Europe, in the form of sending hydro-dewaxing to Akzo, was taking place by 1991; four years later, India was exporting its own technologies and products in this area; and by 2000 it had built global leadership.

Looking back over CSIR's transformation, Dr. Mashelkar judged it to have been accomplished in a responsible manner. He credited the statement of vision and strategy in *CSIR 2001* as having been very critical in creating market orientation while maintaining the quality of science. The adoption of a new management strategy for intellectual property rights in 1996, when CSIR became the first institution in the country to enunciate such a policy, had been important as well.

Meanwhile, numerous changes have taken place in structure and administration: The small projects of the past have given way to large, networked projects, with CSIR's 40 laboratories no longer behaving as independent entities. A strong market orientation has replaced an atmosphere in which work was "individual and group based." Whereas costs had once been no consideration, now time and costs are "sacrosanct" and perfunctory monitoring has given way to stringent monitoring. Similarly, there have been welcome changes in the culture of work: A formerly inward-looking enterprise is now increasingly looking outward, harnessing synergies in all systems. CSIR's outreach has gone from local or national to global, and it currently counts among its partners firms from around the world. Power within the organization itself, traditionally concentrated in Delhi, has been decentralized, with autonomy and operational flexibility replacing the rigid rules and procedures of the past.

To underline his contention that the context decides the content, Dr. Mashelkar turned from the transformation of CSIR India to that of CSIR South Africa, which had been restructured into business units in 1987 and undergone reviews in 1997 and 2002, the more recent of which he had chaired. As the institution had gone strongly commercial, becoming 70 percent self-financing, there had been a significant erosion of its science base. Creating what might be called an "optimum coupling" was very important but always very difficult, he acknowledged. "If you are too strongly coupled with industry, you are always working on today's problems or tomorrow's problems. If you are too far [from industry], you are far from the market and spin off in the wrong direction." And now there was a new challenge facing CSIR South Africa: moving from exclusion to inclusion. Built in the era of apartheid, the institution had to develop a new image, something it was doing rapidly, at the same time that it needed to restore its eroded science base. Even within a given country, therefore, context decided the content.

Dr. Mashelkar concluded with a list of those issues he believed to be key in renewing a national laboratory:

- strategy may change from country to country and from time to time;
- continuous repositioning is essential;
- local relevance and global excellence must be kept in balance;
- "directed" science has to be achieved; and
- creating a "golden triangle" uniting national labs, universities, and industry is crucial.

He added a final word: "Like in every other thing in life, it is all about leadership."

Dr. Atkinson thanked Dr. Mashelkar for his well-organized and articulate description of the transformation that, he said, had clearly brought India's national laboratories from one position in the world to another. He then introduced P. V. Indiresan, a former president of both the Indian National Academy of Engineering and the Institution of Electronics and Telecommunications Engineers and a current member of the State Planning Commission in Delhi. Dr. Indiresan had spent many years with the Indian Institutes of Technology (IITs) in both academic and administrative posts.

NATIONAL AND STATE INVESTMENTS IN SCIENCE AND ENGINEERING EDUCATION

P. V. Indiresan
Indian Institute of Technology (retired)

Dr. Indiresan began with a note of sympathy for his listeners, who, he suspected, might by now be confused as to whether India was doing well or poorly. The contrasting accounts they had heard from earlier speakers called to mind an episode from the Indian epic *Mahabharata* in which the teacher Drona tells one of his students, Duryodhana: "Go into the town and find a good man." Duryodhana comes back saying, "I can't find a good person because everybody is bad in some way or other." Then Drona asks Yudhishthira, Duryodhana's cousin, to "go and find a bad man." And this boy comes back and says: "I can't find a bad man anywhere, because everybody is good in some way or other." Such, Dr. Indiresan stated, was India: Those looking for the bad would find plenty of it; those looking for the good would find there to be plenty of that as well. "What India is going to be, and what India will be as a partner for your own ventures," he observed, "depends, therefore, on you."

Dr. Indiresan's talk would be organized in three parts: what the United States had done for technical education in India, the obstacles India was currently facing, and working to overcome the obstacles in ways that would yield mutual benefits.

The era of modern technical education in India dates to about 150 years before, when the British established three universities and one technical institution in the country. They did this for imperial reasons: They wanted more people trained to work in the colonial administration, and people who "would obey and not think too much." So, rather than copy the Oxford model, they introduced a system of universities whose function was to examine undergraduates and not to do research.

Change took place only after the Second World War, when a formal interest in engineering education coupled with research first emerged in India. For that, gratitude is owed to the United States, and in particular to the late Professor Norman Dahl of MIT, who helped to set up the Indian Institute of Technology (IIT) in Kanpur. Dr. Indiresan recalled from his days as a teacher at the IIT in Delhi the visit there of a British "expert" who told him: "You do the undergrad teaching, we will do the thinking and research, and that will be the partnership we will have." But Dahl, insisting that Indians were capable of research work, introduced a large number of American practices—such basic things as the semester system, continuous evaluation, and the credit system—and encouraged graduate study overall. That spread to the other IITs, which then over time turned out numerous people who had proven outstanding in engineering design.

In the same period, the United States made available through its technology cooperation mission about a thousand scholarships allowing Indians to pursue advanced study in the United States. Unlike today, Dr. Indiresan noted tartly, those who at that time studied abroad returned to India, and it was the recipients of those scholarships who upon their return completely rebuilt Indian science and technology. He therefore reiterated his gratitude for the U.S. contribution to the rejuvenation of science, technology, and technical education in India.

Dr. Indiresan praised the IIT system for its production of excellent designers and analysts, observing that a Wall Street firm looking for an analyst might find that an IIT graduate could do the job better than most. Unfortunately, it has not done as well in innovation, for many reasons, one of which is its low budget—between $20 million and $25 million per year—compared with the billions that Harvard and Stanford have. Still, given that outstanding people have been educated with such scarce resources, India could be viewed either as having done very well or as having been beset with handicaps.

India's caste system presents a second problem. Under one of the ancient models, industry is considered one caste and academia another, and the two do not intersect. One of the few occasions he had had to talk with industrialists was at the present conference, said Dr. Indiresan, adding: "In India, it is hardly possible for an academic like me to meet an industrialist unless a minister is addressing the audience at the time."

A third problem is that of awakening true interest in technology among industrialists. The licensing system that was in effect until 1991 discouraged industry from taking such an interest. That system—"Believe it or not!" he exclaimed—punished a businessman who produced more than a licensed amount: If a business had been licensed to produce 100,000 bicycles and produced 110,000, it was liable to punishment. Things have changed, and quite a few industrialists have now become interested in technology, but some still equate redesign with research even though the two are not identical. And while India is very successful at redesign, research in the true sense of the term has yet to become widespread. "We have got into the habit [of being] content with the technology but dissatisfied with the profit," said Dr. Indiresan, arguing that dissatisfaction with technology would ultimately lead to greater contentment with profit.

A sign of their confusion regarding technology is that Indian industrialists, when they are in the United States or Britain or Germany, visit manufacturers rather than universities. Since they interact only with other industries, they expect IIT to produce what the capital goods manufacturer in the United States produced. "But IITs are not very good at producing capital goods," he said, "although we produce very good human materials."

A problem specific to the IITs was that they are not free to accept donations from abroad, even if the would-be donors are Indians. All donations have to be turned over to the government, which has the absolute right over the funds despite an informal agreement under which the government is to return to the IITs gifts that are intended for them. At the height of the dot-com surge, IIT graduates had come forward with a proposal to invest $1 billion to produce a world-class institution. Dr. Indiresan said that the response of the Indian government—"'you give us the money, and we will decide how to do it'"—had prompted the prospective investors to withdraw their offer.

Taking up the question of the "two Indias," he concurred that there existed both a burgeoning middle class that had become quite prosperous and a very large number of poor people living in conditions "no different from what they were 100 years ago." All were concerned about this enormous disparity. In particular, Dr. Indiresan said, high-quality education has to be made accessible to the poor, in whose ranks could be found extremely brilliant children who could ill afford an education and so had no opportunity to rise.

This concern has recently led the government to propose the "reservation policy," which could be expected to have a negative impact on the IITs. While Dr. Indiresan likened this reservation policy to America's affirmative action in that it is intended to help people who are being left behind, he pointed to what he called a "slight difference": Under affirmative action a person who is not competitive but is competent is admitted, whereas under the reservation policy, he claimed, those who were competitive would be admitted even if they were not competent.

Elaborating, he posited that under affirmative action, as long as a student satisfied an institution's minimum entrance requirements, he or she might be accepted for admission—even without being in the upper echelons of entrants—as a member of a group that had been at a disadvantage and therefore "not had a proper share." Under the reservation policy, in contrast, 49.5 percent of students accepted were to have marks falling below the minimum for admission. How could the quality of an IIT be maintained in this case?

Even leaving this aside, whether the IITs would be able to expand was a major issue. They now admit 4,000 students per year, only 1–1.5 percent of applicants, compared to 10 percent at Harvard or Stanford. For this reason, IIT students are very good—even after four years with teachers like himself, Dr. Indiresan said self-effacingly and with a touch of humor, they emerged unspoiled. Still, certain problems stood in the way of expansion. One of them, of course, is money. The second is the availability of teachers. At the peak of his career, a professor at an IIT receives today the same salary that a new graduate commands as an intern at Dr. Kapur's firm or that of Mrs. Piramal. Finally, Dr. Indiresan noted that a new theory has arisen concerning expansion, according to which simply increasing admissions would produce good-quality people. As he interpreted it, this theory held that "if you cultivate a million goats, you'll get 10,000 sheep out of them."

But perhaps the fundamental problem with the IITs is that, being unique in India, they have no competition. For Dr. Indiresan, the proof was in his efforts to find candidates for admission to the Indian National Academy of Engineering when he was its president. Of 250 faculty members who appeared worthy of consideration, only one was from outside the IIT system—and this in a country with 1,560 engineering colleges. Similar evidence emerged, he said, when the Department of Microtechnology offered to provide funding that was substantial by Indian standards to institutions that would start postgraduate courses. To qualify, an institution had to be engaged in collaboration with industry—and only 11 of the 1,560 engineering colleges met this standard. The lack of competition to which this testified was having a dulling effect on the IITs, he indicated.

That brought Dr. Indiresan to his own recipe for making India a producer not only of good analysts and designers, but also of true innovators: founding private universities. While stressing that he had no complaints about the IITs—he had taught in them for 40 years and found them full of "very nice people"—he maintained that competition for them could not be found within the public sector but had to come from private enterprise. The private institutions supplying this competition, moreover, would have to refrain from accepting government funding. "The government is a very good businessperson," he said, reprising his earlier warning, "and if you take even one single penny, they will extract $1 billion worth of control."

A picture of the government's penchant for keeping institutions of higher education on a short leash emerged from that and other playful remarks that Dr. Indiresan made during the course of his presentation. Whereas the Indian "university system was very tightly controlled and had to do whatever the government wanted," he and his colleagues at the IITs had, he quipped, "enormous freedom: full freedom to do whatever the government says."

Reporting a conversation of that very morning, Dr. Indiresan said that Minister Sibal had asked him what the government should do to maintain standards at the IITs after the reservation policy was inaugurated. "Not that if I tell them, they will listen to me," he commented, "but they are generous enough to ask me for my opinion." Still, he named as his "great advantage" as an Indian the fact that if he were to "say something critical of the Indian government or the Indian system, [his] head [would] not be cut off." Indians, he quipped, "are absolutely free to criticize [anyone] except [their] immediate boss."

Returning to the question of research budgets, he pointed to Research Triangle Park (RTP) in nearby North Carolina as an excellent example of a form of cooperation between industry and academia that India does not have. With its 7,000 acres, 50,000 employees, billions of dollars in output, and hundreds of firms, RTP dwarfs an analogous park planned for India that is to cover 10 acres and involve the "princely" investment of $1 million. The latter project, he said dismissively, amounts to expecting to make a suit from a six-inch length of cloth. What is needed is cooperation that will enable the establishment in the country of an excellent science and technology park that could be not merely self-supporting but a profit center generating income sufficient to support a world-class research university in India. In fact, it constitutes India's "only hope" of meeting the cost of founding and maintaining a research university.

One potential alternative, a scheme for the efficient organization of land called PURA—an acronym derived from "providing urban amenities in rural areas"—had been developed by Dr. Indiresan and had found favor with India's president, A. P. J. Abdul Kalam. While an acre of land could cost as much as $200,000 to $500,000 in an Indian city, just 20 or 30 miles outside the price might be no higher than $10,000. This enormous difference could be used as a tool to fund science and technology parks. Dr. Mashelkar, he said, could explain how to do this, whereas he himself, if asked, was more likely to explain how not to do it.

Summing up, Dr. Indiresan reminded his listeners of the story of Yudhishthira and asked them not to think that India was "very bad" because of what he had said. India was also "very good" and had done extremely well. The country was a world leader in space technology, and, owing to the Green Revolution that began in 1973, it had gone from dependence on agricultural imports to food surplus within 10–12 years and was now the world's leading producer of milk.

Despite its government, which "will do the right thing only after it has tried everything else," he said, "India is a very great country when it sets its mind to be great." Although it might seem to resemble the "young girl in the poem—'When she was good, she was very, very good, but when she was bad, she was horrid'"—Dr. Indiresan warned the audience: "If you ignore India, you will do so at your own peril."

This problem of striking a balance served as the object of a second simile. He likened the innovative mind to a bird: If it is held too tightly, it chokes, but if it is held too loosely, it will fly away. India's government, he observed, tended to hold things very tightly. His hope was that cooperation with the United States would benefit India now just as U.S. advice on organizing technical education had in the mid-20th century, this time both through helping the country establish science and technology parks and through teaching Indians how to hold talented people just right.

Unable, it seems, to resist a humorous story, he used one to present a "warning" regarding the need to keep relations between the two countries in equilibrium. Two partners having purchased a cow, he recounted, one of them said: "You take the front part, and I'll take the back." And it turned out well for him, since his partner had to feed the cow, while he got the milk. In the current case, a more equitable partnership was to be desired.

To underscore this final point, Dr. Indiresan commended the audience to a shloka, or verse, from the *Upanishads*, Vedic texts composed anonymously thousands of years ago:

> *Sahanaavavatu*—Let us both get together.
>
> *Sahanau bhunaktu*—Let us both enjoy together.
>
> *Sahaviryam karvaavahai*—Let us both do great things together.
>
> *Tejasvinah avadheetamastu*—Let great minds flourish.
>
> *Ma vid vishavahai*—Let there be no misunderstandings.
>
> *Om! Shantih, shantih, shantih*—The ultimate is peace, peace, and peace.

Dr. Atkinson thanked Dr. Indiresan for a delightful and enlightening presentation that had not only provided tips on animal husbandry, which he had not expected to glean from the conference, but also had exemplified the value of democracy's support for the freedom of saying, in eloquent terms, what needs to be said.

He then turned to the introduction of his good friend Tom Weber, to whom had fallen the challenge of following such a delightful set of remarks. Dr. Weber had earlier in the year assumed the post of director of the Office of International Science and Engineering (OISE) at the U.S. National Science Foundation (NSF), where he had served for nearly 20 years, more than 10 of them as director of the Materials Research Division.

OPPORTUNITIES FOR U.S.–INDIAN MATERIALS COOPERATION

Thomas A. Weber
National Science Foundation

Dr. Weber said that while he could not promise to match Dr. Indiresan's humor, he could outline some vehicles for collaboration and cooperation and, after providing a little background, tell the story of what NSF had done to further them.

To get started, he projected a statement that the NSF's director, Arden Bement, had made at the Materials World Network Symposium in Cancun, Mexico, in August 2005: "Global collaboration—among scientists, engineers, educators, industry, and governments—can speed the transformation of new knowledge into new products, processes, and services, and in their wake produce new jobs, create wealth, and improve the standard of living and quality of life worldwide."

As was evident, Dr. Bement felt very strongly about the importance of collaboration and the obligation of the United States to promote it. The occasion for his remarks had been the tenth anniversary of a Division of Materials Research program for increasing collaboration in that field between U.S. scientists and their foreign counterparts, a program that had taken some time getting off the ground.

Another reason to look at collaborations was that, as Dr. Weber said with a nod to journalist Thomas Friedman, "the world is flat." Current problems are more complex than those of the past. The number of group proposals received by the Division of Materials Research, according to a study he had just completed, have risen over the preceding few years, a sign that problems were getting harder and that solving them requires teamwork by researchers from various disciplines. The average number of citations was at 10.81 for papers published by researchers at institutions of higher learning in the United Kingdom but jumped to 23.67 for papers resulting from joint work by researchers at academies in both the United Kingdom and the United States. This presumably means that the latter papers are more interesting and better. Similarly, collaboration enables the leveraging of resources that otherwise might not be available to all—data, experience, equipment, and infrastructure among them—even as it furthers the U.S. goal of using its research grants to build a globally engaged scientific community.

But what is the current level of collaboration between U.S. and Indian scientists as indicated by the number of grants? Each of the previous five years had seen between 55 and 80 NSF awards involving joint collaborations. These span all of NSF's disciplines and range in size from several thousand dollars for dissertation research to millions of dollars for major projects. Dr. Weber, often told by the Indian government's Department of Science and Technology (DST) that the number of such collaborations have been decreasing, said an NSF study has shown that the foundation provided $300 million to $400 million annually in overall funding. But NSF's network was a diverse one, with the majority of these moneys going through its research directorates.

His own Office of International Science and Engineering had only $34 million annually, about $10 million of which was spoken for in pass-throughs to the International Council of Scientific Unions, International Institute for Applied Systems Analysis, and other international ventures in which the U.S. government took part. That left his division around $24 million per year to cover international collaboration in all areas of science and engineering, and as a result it cofunded only around 30 percent of total awards.

U.S.–India cooperation, which spans more than four decades, has a fruitful history. Before India became self-sufficient in food production and before the rupee was made convertible, the United States India Fund was set up to fund Indian agencies in rupees. Dr. Weber recalled the time that one of its program directors wanted to give one of his own program directors a bag of money to take back to India, something the latter had been reluctant to do. That initiative had given a start to the NSF–DST joint program, which was geared specifically to those two agencies.

The two nations' relationship continued to flourish during the term of President Clinton, during whose 2000 visit the Indo–U.S. Science and Technology Forum was established, and is flourishing even further under President Bush. A high-level Bi-national Science and Technology Commission is being developed in the wake of the President's visit earlier in 2006, although it was Dr. Weber's understanding that various suggestions had been floated as to what that actually might entail. Briefings were under way at NSF in conjunction with a fall visit to India being planned for Dr. Bement, who likes not only to update himself on scientific research when he goes abroad but also to see what is happening in industry.

Future joint activities might include an institute for advanced study in nanoscience, which Dr. Weber felt would strike those who remembered the old NATO institutes as having potential. Also in the realm of possibility is an advanced institute in geophysics or, in fact, any area of science or engineering. Although some U.S. students currently have the chance to do short-term research at Indian Centers of Excellence, and MIT students were able spend a summer as a research intern at an IIT, the OISE wished to see both more U.S. students visiting India and more Indian students visiting the United States. Under its East Asia and Pacific Summer Institutes (EAPSI) program, NSF is sending U.S. graduate students to Australia, China, Japan, New Zealand, South Korea, or Taiwan for between six and eight weeks, and India could be added to the list.

Dr. Weber then turned to the Materials Research Network, which links U.S. and Indian scientists and is considered the best model for international networking that NSF had yet developed. At NSF, he explained, "materials" runs the gamut from very fundamental research to devices and systems, and it involves multiple disciplines: materials science, physics, chemistry, and engineering. As director of

NSF's Materials Research Division, he was often asked what percentage of U.S. materials research the office funded. He had found it a hard question to answer because, while a chemist can go to the American Chemical Society and a physicist can go to the American Physical Society, materials is spread over at least 20 different disciplinary societies.

The Materials World Network, which had actually gotten off the ground just three years earlier, had its earliest beginning with a series of international regional workshops between 1995 and 2000 that examined ways of stimulating international collaborations in the field. Though held in "nice places," these meetings were not without their frustrations, as evident in Dr. Weber's description of one that took place in Rio when Brazil was on its way to winning the World Cup. NSF then started, over a period of just a few weeks, six International Materials Institutes based at academic institutions in the United States, funding each at between $600,000 and $700,000 per year.

One of these, the International Center for Materials Research (ICMR) at the University of California at Santa Barbara, has very strong ties with India: The Nehru Center in Bangalore ranks as its associate center, and the Indian Institute of Science in Bangalore and the Indian Institute of Technology in Mumbai are also among the 20 member institutions on six continents that form its network. Dr. Weber described the function of ICMR's director, Anthony Cheetham, as "very delocalized" and said that researchers, including graduate students, are traveling back and forth within the network in a variety of exchange programs. ICMR held a summer school in Singapore in August 2005 that featured 35 lecturers and drew 130 participants, as well as coordinating a Singapore meeting of the Asia Materials Network in November of the same year that was attended by researchers from not only India and the United States but also Australia, China, Japan, South Korea, Taiwan, and Singapore itself.

These networks, however, represented only one mechanism of the Materials World Network. Its collaborative projects often begin when investigators from different countries sent their respective funding agencies separate proposals that were then reviewed jointly; if found meritorious, they were funded separately by each agency but in a coordinated manner. Dr. Weber emphasized that proposals did not necessarily come through one central location, and he urged researchers in the audience, particularly those from India, to make their own funding agency aware of their projects and also, perhaps, to submit proposals in order to finance student exchanges.

In the latest Materials World Network competition, one of the nine proposals involving Indian participation to be submitted had come up a winner. As the overall success rate of proposals submitted to NSF's Materials Research Division stands at about 20 percent, and in light of the statistics of small numbers, one in nine is "not that bad," judged Dr. Weber. Cofunded in 2006 by NSF and DST, the winning project, U.S.–India Cooperative Research into Anisotropic Colloidal Magnetic Nanostructures, is seen as contributing to understanding of the physical and biological phenomena that rely on nanoscale magnetic materials. A broader impact is also expected: that of facilitating the exchange of students, at both the undergraduate and graduate levels, and of faculty. Its principal investigators are Vinayak Dravid of Northwestern University and Dhirendra Bahadur of the Mumbai IIT.

As an aside, Dr. Weber remarked that he had been quite impressed by the IIT system and added that the United States has benefited from the fact that a large number of its alumni have occupied all sorts of very high positions here in both industry and academia.

Dr. Weber concluded by laying out obstacles and conflicts that U.S.–India cooperative research activities might face:

- **Data access:** Getting access to geosciences data was very difficult for U.S. scientists.
- **Broadband connection:** India is not connected to the highest-speed broadband research networks, among them the NSF-sponsored GLORIAD, which could be especially helpful to remote collaboration.
- **Intellectual property rights (IPR):** IPR and their enforcement need to be worked out in advance and clearly understood. When U.S. scientists work with the European Commission, there is always a struggle stemming from the universities' refusal to sign agreements that the Commission put forward. This does not represent much of a hurdle in the case of NSF, however, since almost all of its research falls into the "basic" category.
- **Basic versus applied research:** The difference in focus between agencies can sometimes lead to problems.
- **Bottom-up versus top-down selection:** In contrast to agencies that specify the amount of money they are willing to spend in a particular area, an approach he called "top down," NSF typically took a "bottom up" approach, examining proposals that came in over the transom and funding those it considered promising.
- **Distributed versus centralized funding:** Typically, OISE did not receive research proposals directly; rather, its program directors were contacted by NSF research divisions to which proposals with an international component had come in. This was difficult for non-U.S. scientists to understand because many funding agencies around the world, including DST, took a centralized approach to international cooperation. At a conference of the Japan Society for the Promotion of Science (JSPS) that very morning, Dr. Weber had heard the complaint that NSF's international activity had decreased. This was unlikely to be true, he said, but the impression had arisen owing to the lack of a specific NSF–JSPS program.
- **Bureaucracies:** Their influence could be either good or bad, since they had the potential both to accomplish things and to act as a hindrance.
- **Currency convertibility.**

With that, Dr. Weber thanked the audience.

DISCUSSION

Anubha Verma from Georgetown University, remarking that a Nobel Prize in science had last gone to an Indian citizen in 1930, asked when the country could expect its next one.

Dr. Indiresan called the Nobel Prize "an accident" and recommended patience, while also remarking that Amartya Sen had won a Nobel Prize in Economics two years before.

Dr. Mashelkar, praising the question, offered to send along a copy of an article he had written three months earlier explaining how Indians could be the very best in their chosen fields, whether that meant winning the Oscar, winning at Wimbledon, or winning the Nobel Prize. In it, he had analyzed the fact that, in the first 100 years of the Nobel Prize's existence, only three were won by people working in the Third World.

Dr. Newman of CalTech offered the observation that his institution had recorded 32 Nobel "accidents" as a preface to his question, which concerned the role of universities. More and more of the vitality of U.S. innovation is emanating from the universities, whose support has evolved so that it currently comes partly from private sources and partly from diverse government agencies. Suggesting that Dr. Indiresan's bird metaphor, which illustrated the necessity of achieving a balance between control and freedom, might apply to universities and partnerships as well to innovative individuals, he asked what the next steps for India would be and what U.S. academics could do to help.

Dr. Indiresan, warning that it would be difficult to find an answer, cited two of India's chief problems: extremely limited resources, relatively speaking, and fairly limited access to information. The best way for U.S. academics to stimulate and encourage research in India would be to identify bright young faculty members in an Indian university and start collaborative programs with them. This could be of value to the United States as well, given that there were very talented people in India who could not have been as productive as they might be because of a lack of resources.

Dr. Weber stated that the best way to start collaborations was for one person on the U.S. side and one on the Indian side to organize a workshop. That could bring together people who might not have known one another but who, having suddenly found a source of expertise, might want to start collaborating.

Wendy Cieslak of Sandia National Laboratories, describing CSIR's renewal as "amazing" and "impressive," speculated that its scientists must love working in such a vibrant research environment and contributing to industrial and national competitiveness. Piquing her curiosity was the fact that the turnaround chronicled by Dr. Mashelkar, which took place in the 1990s, was followed by a steep rise in publications, citations, and patents beginning in 2000. Had there been a drive to increase the number of publications? Had this rise occurred naturally as the productivity of the laboratory grew? Was it in fact important to Dr. Mashelkar—and, if so, why?

Dr. Mashelkar, recalling CSIR's mission statement, pointed out that in the phrase "scientific industrial R&D" the word "and" appears to have been left out. The omission, however, had been deliberate. In the pre-1991 era, while CSIR did wonderful scientific research, it had no connection with industrial research at all. "There was no competitiveness because we were a closed economy," he said. "Industry produced gums that did not stick, yet people bought them. We produced plugs that did not fit, and we bought them. We produced cars on which everything other than the horn made noise, and we bought them." In this sellers' market there was no innovation, and the bulk of the industrial research that CSIR did at the time was based on reverse engineering.

Change began in the early 1990s: When India liberalized, competition came in, causing industry to change its attitude and look for goods that would be competitive. A story from when Dr. Mashelkar was director of the National Chemical Laboratory would illustrate one of the reasons that India had improved its strength in patenting. Before the economy opened up, whenever he went to industry with work from the lab that was ahead of the rest of the world, he would be asked: "Have *they* done it?" This meant, had it been done in the United States or in Europe?

His response was to reconceptualize the challenge before the lab: NCL would become an international chemical laboratory whose ultimate product, knowledge, could be sold anywhere. "But," he explained, "I couldn't go to General Electric and say, 'I will copy something from you. Will you buy it?' They would kick me out." So the lab had to be basically ahead, and that raised the bar. The patent that he had mentioned earlier—which had, in fact, been licensed to General Electric—had been the subject of a paper in *Macromolecules*, an EC journal.

The effort to couple high science with high technology—to apply the highest level of science to economic, social, and environmental problems—had been the main driver of CSIR's renewal. In most instances when national laboratories are transformed and great stress is put on delivering to industry, incomes go up but the science goes down. CSIR, in contrast, had achieved what Dr. Mashelkar characterized as this "subtle coupling."

Dr. Sinha of Penn State, remarking that the impressive turnaround at CSIR had been accomplished with few changes in personnel, asked Dr. Mashelkar if he could identify elements of the process that might be transferred to other settings with similar success.

Dr. Mashelkar said that, while much had been written by others analyzing CSIR's transformation, "at the end of the day, it is all about leadership." An effort had been made to create leadership within the laboratories by finding people of outstanding merit who would stand tall in science but have a realistic view of the continuity between knowledge and wealth creation. That, he believed, had been the crux of the matter.

Dr. Goel of Nanobiosym, a nanobiotechnology company she had recently started as a research associate in physics at Harvard, asked the panelists to go beyond the discussion of resources to consider the culture of innovation. America had traditionally favored risk taking and innovation; individual initiative and the entrepreneurial spirit were encouraged, and there was a high tolerance for failure. She wished to know what might be done in India, in addition to providing resources and access to education, to create a similar culture.

Dr. Mashelkar said he had visited several universities in Canada before coming to the symposium and learned that, for example, the University of British Columbia had spun off a number of companies, the largest among which had a market capitalization of around $6 billion. Preventing anything comparable from happening in India was a cultural tendency that he represented metaphorically as the division of labor among two Hindu goddesses: Saraswati, the goddess of knowledge, and Lakshmi, the goddess of wealth. "Unlike in this country," he said, "we have never understood the route from Saraswati to Lakshmi."

But cultural change are on its way, if gradually. For the first time in its century of existence, the Ministry of Science was allowing India's scientists to act as entrepreneurs. Suddenly, people such as Ashok Jhunjhunwala were setting up companies like TeNet, or Professor Vijay Chandru Strand Genomics. These examples are of recent vintage, but to multiply them by hundreds—which, according to Dr. Mashelkar, was "entirely possible"—more than cultural change would be needed. Minister Sibal had spoken earlier in the day about an equivalent of the Bayh–Dole Act for Indian universities, as well as about early-stage financing from both the public and private sectors. All this was at its beginning, but liberalization, since it dated only to 1991, was still fairly recent.

Still, risk taking was going to be very critical. Dr. Mashelkar had started a program known as the New Millennium Indian Technology Leadership Initiative, the idea of which to move away from reverse engineering and to work in areas where technologies and markets were not yet established. Funding was in the form of government-backed loans that were to be repaid only in the case of success. The results were, as he put it, "amazing": Within four years a public–private partnership of 65 private-sector companies and more than 20 research institutions had been formed and scored a breakthrough by discovering the first new tuberculosis molecule found since 1963. Several breakthroughs of like significance were on the way, he said, "simply because the government gave us gambling money and a level playing field."

Moreover, the amount of public investment was not overly large, coming to between $10 million and $12 million per year—a sum that, Dr. Mashelkar reminded the audience, nonetheless "goes a long way in India." The entire budget for the country's space program was no more than half a billion dollars per year, although that depended on India's designing, fabricating, and launching its own satellites. It provided such services to Germany and South Korea as well.

Amit Mittal, a science counselor at the U.S. embassy in New Delhi when Norman Dahl and Louis Smullen visited in the mid-1980s, recalled recommending to the Indian government that one of the IITs be increased in size 10-fold.

Mindful of Dr. Indiresan's important observation about the difficulty of diluting the government's dominance of the IITs, he asked whether this was indeed an option.

Dr. Indiresan responded that the IIT Act, which had been inspired by the U.S. system, was the only act of the government of India to grant the kind of autonomy that all educational institutions should have but that no others in India did have. IITs remain, therefore, very special and privileged institutions. He would pass over the reasons that the original plan for the organization, under which it was to depend on the government for only one-third for its means, had not been followed and IIT instead had become 100 percent dependent.

More recently, in 1998, the government declared that the IIT should become as independent as possible and froze the amount of grant funds it was providing. But seeing the IITs' response—they "went merrily ahead and started making a lot of money"—government officials feared losing control and forbade them to accept further outside funding. Until this rule changed, it was going to be very difficult for the IITs to expand because that would require the permission, and the financial support, of the government. Recruiting staff of high quality would be a huge obstacle in any case, as salary levels are restricted by the system.

Dr. Mashelkar said that, in fact, a new picture was emerging with regard to salaries as the entire system of science and engineering education in India was in the process of being overhauled and upgraded. He described this system as a pyramid that had the IITs at the top, followed by the regional engineering colleges, then by the government engineering colleges, with the Industrial Training Institutes referred to by Dr. Kapur at the base.

Recalling that the IITs accepted only 2,000 of 200,000 applicants even though 10,000 others might be as good, Dr. Mashelkar said that the rejects ended up attending regional engineering colleges. In the aftermath of his making a case for giving the regional engineering colleges "a place in the sun," which he had in 1999, they were converted into National Institutes of Technology that received their funding from the central government rather than the states, and their governance structure was changed completely.

In addition, the government engineering colleges and hundreds of other colleges are being lifted up, thanks in part to World Bank support. One of them, the Mumbai University Institute of Chemical Technology, is probably the best chemical engineering school in the country; admission demands minimum marks of 97–98 percent. Dr. Mashelkar, its current chair, said there is a "major transformation" under way. Thanks to the autonomy gained through a public–private partnership, the institute's director was receiving a six-figure salary for the first time ever. He described this as having been done on an "x-plus" basis, with "x" coming from the government and the "plus" from the private sector. He proposed meeting again a year later to see how change was progressing.

Dr. Atkinson offered two observations in closing. The first was that, the previous day, he had had the privilege of lunching with Norman Borlaug on the occasion of the World Food Prize announcement at State Department headquarters. Dr. Borlaug, if he understood the story correctly, had assumed risk in conducting research based on his own commitment to doing something rather revolutionary, something that might make a dramatic change. So in the history of U.S.–India relations, an enormous change had occurred over the past several years thanks to risk taking. If there was a lesson to be drawn from that, it was that risk taking is a very important element.

The second observation was that since the success of a panel was measured in the questions, this had clearly been an exciting panel. Paying his respects to the panel, he added, also amounted to paying homage to Dr. Wessner and the National Academies for putting this conference together.

Building U.S.–Indian Research and Development Cooperation

Moderator:
Mary Good
University of Arkansas at Little Rock

Dr. Good observed that the members of the fourth panel, drawn from the top echelons of research and development (R&D), were in an excellent position to articulate how U.S.–Indian cooperation in R&D could be further strengthened. She then turned to introduce the first speaker, Dr. Swati Piramal, the director of Strategic Alliances and Communications of Nicholas Piramal Indian Limited (NPIL), a major Indian pharmaceutical firm. Dr. Good also noted that Dr. Piramal had been named to the Indian prime minister's board of advisers, and thus had a perspective on discussions of these issues taking place at the highest level in India.

Swati Piramal
Nicholas Piramal India Ltd.

Dr. Piramal said that her talk, titled "Partnerships That Prosper," would present a model for transformation and explain why partnering with Indian companies was good strategy. A few case studies would illustrate the challenges and opportunities arising from the renaissance of science under way in India.

In the quest for the force that reshapes the world, she said, there is always a battlefield; the term she used to refer to it, *Kurukshetra*, signified a battle without which there is no progress. This battlefield was to be found along a pathway or journey to the elixir, which constituted a gift that, in India's case, was the country's science renaissance.

A Transformative Model: "Leadership in Action"

To illustrate, Dr. Piramal laid out a model for "Leadership in Action": After stage 1, a *Call to Adventure*, one would *Cross the Threshold* in stage 2 to arrive at a supreme ordeal on the battlefield, the *Kurukshetra* of which she had spoken. *The Hero's Journey*, stage 3, leads to stage 4, *The Elixir* or *Gift*. The fundamental transformation depicted by this model would be the object of her focus.

Stage 1: Prepare to Journey

This was synonymous with the *Call to Adventure*. It is important here to see the world as full of possibilities, to shift one's view of the world from one of resignation to one of possibility. The speakers of that morning who had pointed to the many things wrong with India might have been correct to do so, but one needs to leave behind the conviction that these things could not change for the belief that they could.

Stage 2: Cross the Threshold

How is this step to be taken? Julius Caesar decided on the 11th of January in 49 B.C. to lead his army across the river separating Gaul, of which he was governor, from the Roman heartland and to undertake a civil war against Pompey, then ruling in Rome. Approaching the Rubicon, Caesar declared: "Once we pass over this little bridge, there will be no business but by the force of arms and dint of sword." Sounding the trumpet, he continued: "Let us march on and go wherever the tokens of the gods and the provocations of our enemies call us." Then he uttered a third and, Dr. Piramal signaled, "very important" sentence: "The die is cast."

Stage 3: The Hero's Journey

"All of us, whether or not we are warriors in the Roman Empire, have this cubic centimeter of chance that pops out in front of our eyes from time to time," she stated. "The difference between an average person and a warrior is that the warrior is aware of this," its being one of the latter's tasks to remain alert to the moment and to act swiftly and powerfully when it arrives.

Stage 4: The Gift

This is the warrior's reward.

"You are what your deep, driving desire is.

As your desire is, so is your will.

As your will is, so is your deed.

As your deed is, so is your destiny,"

said Dr. Primal, quoting from the *Upanishads*. Features of the new reality include the globalization of markets and the intensification of competition that accompanies it and the search for capabilities as technology blurs national boundaries and redefines the value of resources. For India, she added, this new reality includes its own rapidly expanding capability.

President Kalam, an aeronautical engineer, had once urged on India's scientists with the help of this metaphor: "The bumble bee, according to aeronautical design—or models, or mathematics—cannot fly," he said. "But fly it does."

Turning to her own field, Dr. Piramal cited a Cambridge Healthtech report on Globalization of Drug Development published in June 2006 that, surveying 235 executives in the pharmaceutical industry, found that three-quarters of their firms were engaged in drug development in India. Leaving behind its days as a destination for clinical trials, "low IP sensitive discovery outsourcing," and custom synthesis, India now offers high-value services and partnering opportunities. The skill gap is contracting quickly, and technology is helping India leapfrog its competition and create strategic alliances and other linkages based on either equity or core capabilities. The report held that India currently presents a greater opportunity for improving Western R&D productivity than did China.

Partnerships: Many Objectives, Multiple Models

Partnership objectives are many: risk sharing, economies of scale, market-segment access, technology access, and, as Minister Sibal had noted earlier, geographic access. There are multiple partnership models as well; Dr. Piramal posted a chart on which possibilities were arranged in function of the partners' degrees of commitment and ownership (see Figure 5). She stressed that the only true partnership was a "win-win partnership" in which all parties share both risk and reward equally.

Illustrating the Benefits of Partnerships

Dr. Piramal then offered as an illustration her own company, NPIL, a "practitioner of Indo–U.S. cooperation" that on the strength of partnerships with many American firms had risen from its 1988 ranking of forty-eighth in the formulations area of India's pharmaceutical sector to fourth in 2005. NPIL's pursuit of growth through strategic acquisitions, alliances, and joint ventures had brought it into partnership with many foreign firms, among them Allergan, Aventis, and

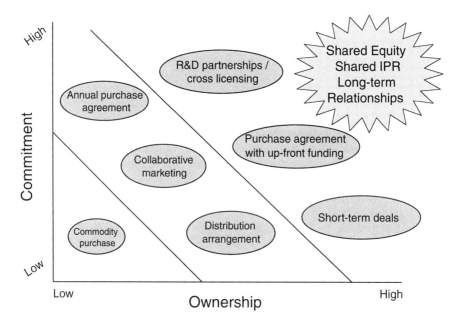

FIGURE 5 Partnership Models.

Roche. NPIL had put into practice its stated mission—"making a difference to the quality of life by reducing the burden of disease"—on the Indian subcontinent, adding value across the drug-discovery and development processes, whether via R&D, contract manufacturing, or clinical research.

Meanwhile, it had acquired a global footprint by extending its reach into the United Kingdom, the United States, Canada, and parts of China. Passing through U.S. Customs that very morning, Dr. Piramal told the agent she had come to look after the affairs of her company and added that it employed around 1,000 Americans, who made glass vials at three manufacturing facilities. Remarking that the company's research scientists were of 22 nationalities, she said: "This is not only about a reverse brain drain, it is not only about outsourcing, it is really about acquisition of global talent."

Just the day before, NPIL had signed an agreement to acquire from Pfizer a 450-employee facility at Morpeth in the United Kingdom in a transaction carrying a supply agreement that was good through October 2011 and could potentially yield upward of $350 million in revenues. This, Dr. Piramal said, amounts to changing the game: Involved were patented products, not low-value generics, and the company would be supplying them to over 100 countries. Since entering the custom-synthesis market in 2003, NPIL had already taken its place among the world's top 10 companies, having described "a trajectory of change from nothing to becoming a global powerhouse of talent, particularly in the areas of chemistry and custom synthesis."

Scientific "Horsepower" at Lower Cost

In 2005 the company had opened an R&D facility, and Robert Armstrong of Eli Lilly, the next speaker, had recently brought a team of more than 30 to visit it and discuss partnering. Even if NPIL spends what some considered too little on research, Dr. Piramal said, it is able to buy "a lot of scientific horsepower" for its money and so far in 2006 had filed 14 New Chemical Entity (NCE) patents, granted on compounds that are entirely new rather than variants of previous patents. Of these, one was already in clinical trials and three others were about to enter that phase.

NPIL had become the first of India's private firms to participate in an expedition to Antarctica as part of an effort to mine both Antarctic and oceanic microbial diversity for biotechnological applications.

The company's pipeline is rich, and its dream is to develop a new drug for $50 million with the potential to go global—an effort that, she said, could cost up to $1 billion in the United States. Even if her cost estimate proves to be off by 100 percent, she said, the gap between $100 million and $1 billion is still significant.

President Clinton had visited NPIL in 2000, and one of the things Dr. Piramal told him was that India lacked good patent attorneys and patent examiners. Since then, India's backlog of patents has shrunk from 22,000 to 6,000, the result of efforts by both the government and private sector to stimulate change.

Best Practices for Building Partnerships

She then offered a rundown of best practices for building partnerships:

- anticipating business risk, which includes making a good business plan internally and fostering openness so that it can be refined through interaction;
- understanding rights and obligations, under which she placed reducing complexity, "playing for the long run," and eschewing short cuts;
- preparing realistic feasibility studies rather than overpromising and paying for it later, an important point for Indians, who "always like to promise a lot";
- defining expectations clearly;
- rewarding performance;
- finding the best talent; and
- creating planning to bridge management styles.

Elaborating on the last point, Dr. Piramal noted that U.S. firms evince a greater desire for order than did their Indian partners. Because of India's chaotic infrastructure, people there rely on "a very quick way of thinking" to achieve "excellence in chaos." She quoted the famed Sam Pitroda to the effect that in India, "if you drive a car, you will find cars coming at you in all directions—but, somehow, we get there." At ease operating in these conditions, Indians find that, in contrast, Americans like structure, plans, and systems—and this difference needs to be managed.

A List of Partnering Mistakes

Complementing the list of best practices for partnership was a list of partnering mistakes:

- being a "possessive child," Dr. Piramal's label for an attitude characterized by everything from excessive protection of one's own technology to erratic communication with one's partner;
- lacking trust;
- failing to attract the best people;
- picking the wrong "spouse";
- being vague about objectives and goals; and
- stifling of an alliance's growth by its "parents."

Dr. Piramal then put forward a "Formula for Success" tailored specifically to undertaking partnerships in India. "Look before you leap" was a watchword: Indians' aversion to the legal process, and their consequent adherence to the spirit as well as the letter of the law, underscored the importance of due diligence. "When you sign the agreement, you throw it away and really live for the spirit of it," she explained. "We believe that legal fights are long and unpleasant and that business, like everything else in life, should be a joy." Once a relationship is formed, it should be built gradually, with the partners focusing on common ground and shared goals. And since relationships can be reshaped over time, whether due to the exit of an original partner or a change in thinking that arrives with a new manager, structural adaptability is very helpful. "Being a farmer, not a hunter-gatherer," an apparent recommendation to take the longer view, was the final element in her formula.

Allergan India: A Case of Indo–U.S. Collaboration

For a case study, Dr. Piramal turned to Allergan India, a 10-year-old joint venture of NPIL and the Orange County, California, company Allergan Inc. that had been nominated for the 2006 "Best Partnership Alliance" award by *Scrip Magazine*. This arrangement, which features "the leading Indian player working with the leading U.S. player" in ophthalmic eye care, has a strong foundation: The partners selected each other for values, purpose, and complementary strengths. Even while taking advantage of mutual trust and credibility and of each other's capacities—in technology and research, among others—to build, the partners "reinvented" their relationship every year through exchanging managers and expanding the joint venture's agenda. The chairman of Allergan Inc., David Pyott, had called Allergan India "an excellent model demonstrating the benefits of collaboration between a global MNC and a strong Indian company."

But was this success replicable? Could such alliances be the source of more earnings, and could they be depended upon for more growth? Dr. Piramal saw the keys as knowledge sharing, establishment of knowledge management, capture and dissemination of best practices, and alliance training. NPIL itself repeated its Allergan success with a second California-based company, Advanced Medical Optics, to build one of the largest custom-manufacturing relationships in India, as well as with two firms in the biotechnology sector, Gilead and Biogen-Idec.

Returning to her "Leadership in Action" model, Dr. Piramal said that what made the *Call to Adventure* of Indo–U.S. partnerships so interesting was the transformations they would further in both nations. Calling partnerships, particularly in joint development, a "strategic imperative for most firms" in the United States, she predicted that Indo–U.S. cooperation in the life sciences would increase.

To conclude, she screened a short video called "The Science Anthem" reflecting the renaissance of science in India.

Dr. Good thanked Dr. Piramal and praised her video both for its beauty and its potential for use in other venues. She then introduced Robert Armstrong, Senior Vice President for Discovery Chemistry Research and Technologies and Global External Research and Technologies at Eli Lilly and Co.—a firm that, she noted, has extensive interactions with India.

Robert Armstrong
Eli Lilly and Company

Thanking Drs. Wessner and Good for inviting him to speak, Dr. Armstrong said he would guide the audience through Eli Lilly's approach to the strategic challenges facing U.S. business in recent years as it operated in the innovation space in Asia. Before taking this up, however, he declared his wish to warn against underestimating a company's need to develop collaboration and understanding of its business internally as a prelude to reaching out and setting up external collaborations.

Being There: The Need for Acquaintanceship

To begin, Dr. Armstrong addressed under the rubric of "Communication in a Rapidly Changing Environment" the importance of comprehending the current dynamics of movement in Asia and, specifically, in India. He was, he said, forever asking himself: "How often should I be on the ground in India?" There were new buildings and new products and new companies appearing all the time in this "very, very entrepreneurial environment"—an environment that, he believed, many in the pharmaceutical industry has not yet fully understood.

He recalled as a turning point for Eli Lilly a meeting of the company's R&D executives that had taken place around two-and-a-half years earlier. Having been asked to give a presentation on Asia strategy, Dr. Armstrong arrived with a deck of 20 to 25 slides. The group made it no further than slide No. 1, however, owing to its lack of knowledge of issues ranging from the state of Indian science and the country's intellectual-property regime to government funding for its labs and graduation rates for the scientists who would be feeding personnel pools for some innovation activities. Drawing what he felt to be the obvious conclusion, Dr. Armstrong turned himself into a travel agent and, for the next six months, worked to get his colleagues on the ground in India. That allowed Lilly's executives not only to arrive at a degree of internal alignment—which was transferred to some actionable items in ways that he would explain—but also to speak with a single voice as they started to visit potential Indian partners.

The benefit was apparent at an R&D evening Eli Lilly held for pharmaceutical industry executives in 2005 that was attended by Dr. Piramal along with representatives of a number of other Indian firms. For the first time, according to Dr. Armstrong, the two sides began to see paths to innovation they could go down together that might allow them to leapfrog the more linear, traditional way of doing R&D. He would go into some of the details later on, then conclude by explaining how to put the "R" into "R&D."

Vaulting Barriers to Productivity Growth

Displaying an April 2002 article from the *Wall Street Journal* with the headline "Why Drug Makers Are Failing in Search for New Blockbusters," Dr. Armstrong remarked on how frequently one saw a version of the accompanying chart, which showed drug development costs increasing as the number of NCEs remained fairly flat. To prove the point, he posted a *New York Times* article, this one from January 2006, with a similar illustration.

Eli Lilly's efforts to dissect and solve this problem had yielded a productivity equation that the company is using internally in which P represents "pipeline"; *WIP*, "work in progress"; $p(TS)$, "probability of technical success"; and *CT*, "cycle time":

$$P = \frac{(WIP)(p(TS))}{(\text{Value})} \rightarrow \downarrow \cos + \downarrow c + \uparrow \underline{p(TS)} = \uparrow \text{Productivity}$$

The equation, Dr. Armstrong explained, was a mechanism enabling a company to look at its "very large" pipeline and to achieve an understanding of whether it was full in accordance with historical expectations for delivery and flow or instead contained gaps—and, in case gaps existed, to figure out what might be needed to fill them. Embedded in the numerator, along with the probability of technical success, was a *Value* component attached to the disease state and modality in question: The latter could be aimed either at ameliorating some of the side effects of the former or at going after its fundamental causes. The denominator, in addition to cycle time, contained a component reflecting the *Cost* of running the entire drill.

India's Role in Improving Productivity

Although every term of the equation has relevance for increasing productivity, Dr. Armstrong chose to focus on those that might elucidate Eli Lilly's thinking on the issue and, in particular, on the role it saw India playing. All the activities of the pharmaceutical industry's R&D sector can be broken down into two components and placed on two axes—the x-axis representing technical difficulty on a continuum from hard to easy, the y-axis ownership on a continuum from proprietary to nonproprietary—so that the origin of the axes marked the confluence of the highest difficulty with the greatest ownership (see Figure 6).

Among the factors determining the position of research activity along the x-axis might be coordination of multiple scientific and medical backgrounds, level of experience in a particular area, and the status of activities that might

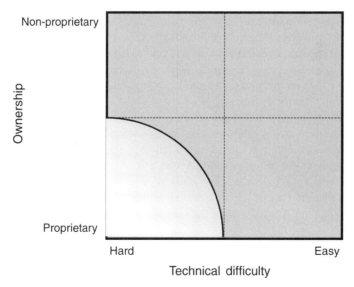

FIGURE 6 Implications of project difficulty and know-how on external partnering arrangements.

have been very innovative at one time but had become industrialized to the point of having converted to standard operating procedures. Dr. Armstrong's comment regarding the *y*-axis was that India's embrace of intellectual property rights protection both enables multinationals to consider undertaking innovative activity there and created value for local industry. As evidence of the latter point he offered the investment in innovation that is beginning to be made by such companies as NPIL to target diseases that affect India's region and thereby meet the medical needs of its population.

Running through Eli Lilly's analysis, Dr. Armstrong noted that pharmaceutical industry R&D activities tended to be judgment-based at their debut but evolve over time so that they became rule-based. The question before the company was which levers in a productivity equation could be affected so as to ameliorate such factors as cost and cycle time. The industry, motivated largely by a desire to drive down the former, has so far considered outsourcing for what it perceives as rules-based activities and, in his opinion, has been successful in so doing. However, what has been eye-opening for the Western pharmaceutical industry—even if it was hardly surprising to the companies providing the services—is that a "huge" component of cycle-time reduction has also been involved. "Embedded in that cost reduction is a lot of innovation in processes and focus in delivering products," he explained, "so that outsourcing or outlicensing activities previously occurring inside the company has changed the equation on cycle time for many of them."

Spreading Knowledge Within the Company

The degree of appreciation for this relationship varied even within Eli Lilly, as some who attended the meeting of R&D executives previously mentioned had already engaged in collaborations with India. Like all global pharmaceutical majors, Lilly had been active in India for years, teaming with a number of companies there both on the manufacture of many legacy products that still carried its own brand and on the launch of global clinical studies. So, some groups within the company were quite familiar with the concept of outsourcing to increase productivity and had seen it validated in India. But others, especially those involved in early development, remained hesitant about a global model that called for transferring next-generation activity abroad and, in general, embracing innovation wherever it might occur. The company was, however, making substantial inroads in that direction.

Alliances: Not Only Cost-Cutting Vehicles

Over the previous decade, in fact, research alliances had been Eli Lilly's primary mechanism for addressing cycle times, and the company had benefited from process changes and technology achievements made by some of its partners. But in addition, Dr. Armstrong stressed, alliances are becoming vehicles for improving the probability of technical success.

In quest of the latter, Eli Lilly has invested in the firms with which it is allied, not only in the area of manufacturing innovation but also, as he put it, "in true innovation, at the front end of the pipeline feeding the R&D engine." To date, these research alliances have been with biotech companies in the Boston, Seattle, and San Francisco areas; around Cambridge and Oxford in the United Kingdom; and in Germany.

But on their continuing visits to Asia, and in particular to India, Eli Lilly officials were being reminded by the numerous start-ups they were seeing of what they had observed a decade before in San Diego and five years before that in San Francisco and Boston. Beyond mirroring the entrepreneurial achievements of those areas, where the risk taking of start-up companies was often quite well financed, Indian start-ups were taking on problems that lent themselves to "leap-frog solutions."

A saying current at Eli Lilly—"we do discovery and development without walls"—is intended to convey that the company could not expect to do everything on its own. Finding partners it could work with is therefore regarded as key, and Dr. Armstrong and his colleagues see India as providing global pharmaceutical companies in the coming years with "very fertile ground" for true collaboration, particularly in the innovation space.

Simultaneously with a trend in the sector toward moving activity out when that might get it done in a more time-efficient fashion or at a lower cost, an opposite and perhaps more important trend is emerging: Partnerships are increasingly in evidence at the early stages of innovation. Not only are companies in the R&D services or materials procurement sectors "moving their way into the discovery component," but stand-alone biotechnology firms whose mission at start-up has been to operate in highly proprietary and technically difficult areas are also forming alliances. Eli Lilly is spending a great deal of time assessing companies worldwide in an effort to identify those it might want to collaborate with on pilots or experiments, and Dr. Armstrong predicted that, as the sphere of partnership activities develops and matures, Lilly would be able to pursue productivity improvement by tapping into these partnerships more effectively.

A "Mosaic of Innovators" in Pharma's Future

To help illustrate how the changing distribution of skill around the world is affecting the way the industry does business, Dr. Armstrong displayed a chart (Figure 7) highlighting a number of chemical starting points and some of the issues surrounding them.

There were many companies in Asia, including in India, "that map very nicely to a large number of components of how we think the next generation of chemical starting points is going to occur in the pharmaceutical industry," he said.

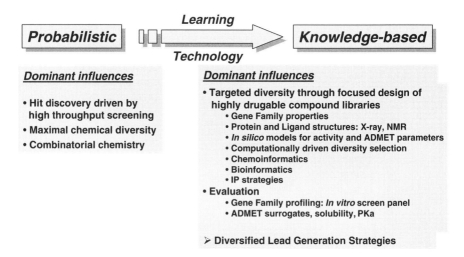

FIGURE 7 Example: Identification of chemical starting points.

In keeping with the evolution of drug discovery at Eli Lilly, the company intends to put together a "mosaic of different innovators" that would allow it to address productivity improvement throughout the process of innovation.

To conclude, Dr. Armstrong stressed how fortunate Eli Lilly has been in having conversations at a very high level with a large segment of India's pharmaceutical industry and in achieving a thorough grasp of both the opportunities and the challenges facing the companies that these Indian companies represent. At the same time, he and his colleagues have been grappling with a challenge of their own: trying to achieve an alignment inside Lilly that would better enable it to work in the innovation space on a global basis.

Dr. Good, offering Dr. Armstrong enthusiastic thanks for his presentation, presaged the following introduction with the observation that, while 25 years before the word *global* was scarcely to be seen in the corporate title of an R&D official, it currently figured in most of them. This was exemplified by the next speaker, Kenneth Herd, who had recently been named Global Technology Leader for the Materials Systems Technologies at GE Global Research.

Kenneth G. Herd
General Electric

Outlining his talk, which would focus on GE's efforts to nurture cooperative ventures in R&D bridging the United States and India, Dr. Herd said he would begin with a brief overview of GE and of its Global Research organization. He would then turn to GE's activities in India, both past and present, before specifically reviewing GE's involvement in R&D there and concluding with a discussion of some of its achievements, opportunities, and challenges.

GE is organized into six businesses serving customers at the industry, market, and national levels, which Dr. Herd listed with their focuses:

- **Health care:** diagnostic imaging, clinical systems, information technology, services, biosciences;
- **Infrastructure:** aviation, energy, rail, water, oil and gas;
- **Industrial:** consumer, plastics, silicones, security and sensing, equipment services;
- **Commercial finance** and **consumer finance:** insurance, leasing, financial services globally; and
- **NBC Universal:** TV and radio networks and stations, entertainment, sports.

With operations in more than 100 countries, GE has a workforce exceeding 300,000 and, for 2005, earnings of around $16.6 billion on revenues of $155.4 billion. "We are very proud to be the only one of the Dow Jones Index's six original companies that is still listed on it," he said.

The Evolution of Research at GE

GE Global Research has evolved from one of the first industrial research labs in America into a highly centralized R&D organization that is among the world's most diverse. Begun in 1900 in a barn behind the Schenectady, New York, home of Charles Steinmetz—a brilliant GE electrical engineer who recognized R&D's critical importance to the development of commercial products—it has since grown steadily into a truly global presence. The John F. Welch Technology Center in Bangalore, to which Dr. Herd would return in a moment, became GE Global Research's first technology center outside the United States when it opened in 2000. Its China Technology Center, opened in 2003 in Shanghai, currently employed about 1,200; GE Global Research–Europe, which opened the next year in Munich, had around 100 employees.

GE in India: A Long History

GE's business experience in India began when it installed the country's first hydropower plant in 1902, and all GE businesses have been present since 1998, representing a wide range of activities in manufacturing, services, and technologies. The company has more than 12,000 full-time employees and $2 billion in assets in India, and, in 2005, posted revenues of $1.4 billion while exporting $1 billion in products and services. Just two weeks previously, on May 30, 2006, GE Chairman Jeff Immelt had announced in a speech to the Bombay Chamber of

Commerce that the company would invest $250 million in infrastructure, health care, and real estate in India. Its current goal was to acquire $8 billion in assets in India and reach an annual level of $8 billion in revenues there by 2010 through activity in health care, cleaner energy, clean water, and aviation.

Today's Attraction: Intellectual Capital

Most compelling among the reasons for GE to develop R&D capability in India is the country's strong intellectual capital, as embodied specifically in its talent pool of engineers and scientists. India's 200-plus national laboratories and 1,300 industrial-sector R&D units employ approximately 200,000, and, owing to India's more than 300 universities, the student pipeline is very strong as well. An R&D presence enables GE to attract India's best global talent and apply it to critical technology development in existing and emerging markets. In return, GE is in a position to invest in training and developing world-class talent in India and to offer jobs that would attract great Indian talent back from abroad.

GE's Bangalore Technology Center

GE's technology center at Bangalore, an $80 million investment, comprises 500,000 square feet of facilities that are spread across a 50-acre campus. Employing approximately 2,500 engineers and scientists—over 60 percent of them with advanced degrees and 20 percent with global experience—its state-of-the-art laboratories conduct research and development in many disciplines: mechanical engineering, electronic and electrical system technology, ceramics and metallurgy, catalysis and advanced chemistry, chemical engineering and polymer science, new synthetic materials, process modeling and simulation, and power electronics and analysis. Based there in addition to a Global Research technology team are teams from other GE organizations: Healthcare, Plastics, Silicones, Water Technology, Energy, Consumer & Industrial, Aviation, and Rail. These groups of engineers and the technologists work together to transition and deliver technologies out to GE's businesses.

To date, the Bangalore center had filed over 370 patents, 44 of which had been issued, something the company is very proud to have achieved in less than five years. Among the center's recent technology successes:

- **GE Healthcare:** diagnostic-imaging breakthroughs for computed-tomography and magnetic resonance products;
- **GE sensing:** phased-array ultrasound system developments for inspecting critical aircraft components, as well as next-generation pressure sensors; and
- **GE Plastics:** high-performance plastics solutions for the automotive sector.

GE Technology in India: Four Phases of Development

The development of GE's technology efforts in India can be viewed as having gone through four distinct phases:

1. a period of contract engineering during which the predominant activity was the outsourcing of engineering analysis and modeling tasks to non-GE businesses in India;

2. the decision to establish a foothold in India by building a solid technology foundation there, which led to the engagement of senior leadership across the corporation in investing in the John F. Welch Technology Centre as a key resource. Dr. Herd acknowledged the "significant support" GE had received from one of the day's earlier speakers—R. A. Mashelkar, the director general of the Council on Scientific Industrial Research—in the center's start-up and launch;

3. the process of recruiting and staffing the center with "the world's best," which had begun with identifying the centers of excellence that would reside in Bangalore, progressed to hiring a mix of experienced and recent graduates across a range of disciplines, and continued with significant investment in the training and mentoring of the new teams, which were active along a wide spectrum of global technology programs in vital areas; and

4. reaping the fruit of these collective efforts, the phase that is currently in progress. "Unique intellectual property and design concepts are being generated, and the infrastructure and processes are in place to drive GE's strategy of growth through technology," he said, describing the company as also very well positioned to develop technology for the markets emerging in India and the region in aviation, energy, rail, health care, consumer products, and water technology.

As with any new venture, the launch of GE's global R&D programs has faced challenges. While there was much concern at first about work shifting out of its U.S. labs, Dr. Herd recalled, GE has continued to strengthen them while significantly expanding its collective technical breadth by growing its labs abroad—without whose addition, he maintained, GE would have been unable to keep up with its growing technology needs.

Lessons Learned from GE's Indian Experience

He concluded with a list of lessons that GE had learned from its involvement in India:

• **that global teaming is not intuitive** but requires training, tools, and a cultural shift whereby both sides meet in the middle. "In fact," said Dr. Herd, "global teaming has added a whole new dimension to our careers—the experience of bridging global divides, exploring new cultures, addressing global challenges—and continues to be a very enriching experience for all involved."

• **that schedule flexibility is very important** to accommodating time-zone differences. Dynamic work hours, global video conferencing, early-morning and late-night teleconferences, and global travel had become an integral part of GE's routine.

• **that linkages among sites are absolutely critical.** GE had found that by first defining core competencies or centers of excellence at its global labs, then linking them to adjacent centers of excellence in the United States, it could avoid redundancy and competition among its labs abroad while leveraging the experience and network of its existing domestic teams.

• **that retention correlates with the vitality of work.** When the global teams are energized by their work, they have a direct line of sight to its product applications, and retention is high. In contrast, when work is of lower quality, vitality, and visibility, retention is understandably poor. GE's highest priority was building high-impact, high-visibility programs at its global sites, an important challenge these sites were in the process of taking on.

• **that start-ups take time.** Building and training the teams, making the organizational and cultural shifts, and "hanging in there" have really paid off, said Dr. Herd. Equality of skills, visibility, engagement, and vitality of innovation were becoming reality. As a researcher and technology leader at GE, he had seen personally that the R&D landscape is accelerating rapidly both in India and on a global scale. "We are very fortunate to be a part of this very exciting frontier," he concluded.

Dr. Good, thanking Dr. Herd, reflected that while each firm had found its own structure for partnership, all were improving their business positions—an indication that the diverse structures were working. "That's probably as straightforward as we can get," she remarked, before introducing Ponani Gopalakrishnan, former Director of the IBM India Research Laboratories, as a man with a great many friends in both the New York IBM community and Washington.

Ponani S. Gopalakrishnan
International Business Machines

Dr. Gopalakrishna expressed his gratitude to the audience for staying so late in the afternoon and pledged to do his share by trying to measure up to the level of the brilliant and informative talks that had preceded his. He would offer a brief view of IBM's experience in running what was probably "one of the only information technology pure research organizations in India," which was founded in 1998 and pursued technologies for the next generation of the company's products and services.

Investment to Increase IBM's Indian Presence

The organization that Dr. Gopalakrishnan's headed is part of an IBM "fairly strong" presence in India. This presence is likely to grow stronger still as the result of a planned $6 billion investment in the country that IBM made public only 10 days before the day of the National Academies' conference. In conjunction with that announcement, President A. P. J. Abdul Kalam addressed a gathering of about 10,000 IBM employees in Bangalore, while 20,000 of their coworkers in cities such as New Delhi, Mumbai, Hyderabad, and Calcutta saw him speak via videoconference.

IBM India spans not only these cities but also a spectrum of functions, from software and systems development to the pure, long-term research done by the organization Dr. Gopalakrishnan headed. IBM's R&D personnel in India number more than a couple of thousand, in addition to whom there was a large contingent of employees delivering information technology services and IT-associated services to a worldwide client population.

Profiling IBM R&D in India

IBM India Research Laboratories itself has a little over 100 very highly skilled researchers working as part of a team of about 3,200 that is spread over the IBM Research Division's eight labs worldwide. The lab in India has the same mission as many of the others—to advance the state of the art in information technology—while working primarily on software and services-related research, and taking into account local and global priorities alike. Its essential focus, however, is to provide leadership for the future both to the IBM operations in the region and to the company's clients and partners both in India and globally.

IBM India Research Laboratories has its headquarters in New Delhi and a team in Bangalore conducting research directed toward the company's services organizations. Dr. Gopalakrishnan named key areas in which the labs are engaged, cautioning that the list is representative rather than exhaustive:

• **Information management.** Enterprises increasingly find that internal data are no longer stored exclusively in regular relational databases but in multiple forms that include e-mails, text messages, PowerPoint presentations, and the like. The problem of capturing the information and gathering intelligence from these varied data is being addressed through research into the integration of information from structured and unstructured sources.

• **Software engineering**. Tools, methodologies, process optimization, and modeling were among the fields of inquiry.

• **User-interaction technologies.** Work is in progress on speech-recognition technology applying to the local languages of India, Hindi, and Indian-accented English—the latter's being, for this purpose, "a different language," Dr. Gopalakrishnan said.

• **Services research.** Begun in Bangalore late in 2005, this work is being carried on in conjunction with that of IBM's services delivery teams and was focusing in particular on issues arising in IT-enabled services. He called the activity "pioneering" in that R&D has traditionally been associated with products but a "strong component" of the technology involved in product research could be applied to the services as well, an endeavor whose potential IBM is starting to investigate. Among the areas being looked at by the Bangalore team are process modeling, system resiliency, knowledge-management platforms, and infrastructures for maintaining knowledge-workforce management. A paradigm similar to that applied to a parts supply chain might be applied to human-skills management as long as completely different methodologies and approaches to modeling were taken.

This work is being pursued in close collaboration with what he referred to as IBM's "global technical ecosystem," which includes not only the company's seven other research labs but also its partners elsewhere in industry and in academia.

He then showed a brief video designed to portray the scope of IBM India Research Laboratories' efforts in these domains. The video focused in part on a current project in which the labs were laying foundations for a very secure network conceived in accordance with the realities of India's information technology infrastructure. Groundwork for this project began with a "fairly thorough analysis" of the information-sharing needs of India's health-care entities, which range from large hospitals with very sophisticated IT systems to 10-bed clinics run by a pair of physicians that might have no IT system at all. The network would attempt to take into account such diverse requirements.

IBM's Four "Secrets of Success"

Reflecting on his experience, Dr. Gopalakrishnan raised the question of whether IBM could expect its technological research and innovation in India to match the level and quality of that conducted in many of the research labs it operated elsewhere. "The answer is a definite 'yes,'" he declared, imparting four "secrets of success":

1. Address the right problem. In an industrial research lab, this meant making sure researchers focus on the problems that would eventually generate the broadest business or social impact, a point he would return to in No. 4.

2. Build the right skills. IBM India Research Laboratories works very hard to attract and retain talent with wide international experience—talent equal to that at any other IBM research lab. Fully 50 percent of its researchers had Ph.D.s and, of those, 50 percent had earned them in the United States; Dr. Gopalakrishnan himself was a product of the University of Maryland–College Park. The permanent research staff is supplemented through a regular program of academic visitors who stay for between three months and a year and through internships offered to university students from countries around the world, including the United States, Australia, and India itself.

3. Encourage collaboration. Many of the lab's projects involve collaboration with researchers, both within IBM and outside, from other parts of the world.

4. Ensure broad impact. The practical significance of their work is what gives researchers the satisfaction that keeps them motivated.

India has certain unique market requirements that make for very interesting design points, Dr. Gopalakrishnan said, one example being seen in the fact that it offers the lowest-cost GSM cellular phone service in the world. Invoking C. K. Prahalad's book *The Fortune at the Bottom of the Pyramid: Eradicating Poverty Through Profits*, he noted that there are compelling arguments for designing systems and solutions for consumers at the lower end of the market—not for philanthropic or charitable reasons, but because real money is to be made there. An analysis of affordability and utilization suggests interesting technologies that could be used to develop systems that allow very high volume access at low cost. IBM India Research Laboratories is itself embarking on a project aimed at providing information technology to the lower end of the economic pyramid.

Potential Areas for Expanded Indo–U.S. Collaboration

Considering potential areas for expanded collaboration, Dr. Gopalakrishnan speculated that the extensive technology base embedded in the U.S. universities and in its other research organizations might serve as a source upon which Indian industry could draw as it addressed the challenge of taking on international scope. Great value might also be found, as all tried to deal with what was a growing pool of research, in coming together to focus on some grand challenges. He put forward as an example the aforementioned health care data network: India's greenfield environment holds the potential for experimenting with technological capabilities and possibilities that could be applied to U.S. systems as well.

In conclusion, Dr. Gopalakrishnan offered his own opinion that there are significant opportunities for collaboration currently available. He also drew attention to the remarks made 10 days earlier in Bangalore by President Abdul Kalam, in which he outlined a compelling vision: "the creation of a World Knowledge Platform for realization of world-class products for commercial applications using the core competencies of partner countries which will meet the needs of many nations. . . . Initially, the mission of [the] World Knowledge Platform is to connect and network the R&D institutions, Universities and Industries. . . ."

Dr. Good then introduced the day's last speaker, M. P. Chugh of Tata Auto-Comp Systems (TACO), who is based currently in Troy, Michigan. She noted that Tata, TACO's parent, is one of India's largest diversified companies.

M. P. Chugh
Tata AutoComp Systems

At the risk of disappointing the audience, Mr. Chugh said, he would admit that he was neither a scientist nor an R&D professional but a "simple business-man" preoccupied with return on investment. Then, recalling a statement made earlier in the day by Dr. Kapur about the importance of "return on innovation," he raised the question of what innovation really is. According to Tom Peters, innova-tion is never created in institutions, but always by breaking the rules. Offering the examples of Google and Microsoft, which had broken many of their industries' rules—or had, at least, violated many conventions—Mr. Chugh reiterated his uncertainty about the nature of innovation and of R&D, as well as about the role institutions played in them.

After profiling the 29-company Tata Group and the unit he himself repre-sented, Tata Auto Components (TACO), Mr. Chugh said that he would enumerate the lessons learned from U.S.–Indian cooperation on what he called "develop-ment and engineering," drawing mainly on the group's involvement in Tata Johnson Controls. He would then touch briefly upon an ongoing U.S.–Indian R&D project with which he had been associated and, to conclude, address the opportunities that Tata believed were offered by cooperation.

The Tata Group: India's General Electric

Known to some as the "GE of India," the Tata Group is the source of 5.1 per-cent of the country's exports and 2.8 percent of its GDP. One of India's largest con-glomerates with interests in the automotive, steel, power, chemical, information technology, tea, coffee, and hotel sectors, it employs 215,000 workers and posted turnover of $17.8 billion in its 2005 fiscal year. Its total market capitalization of $41.4 billion was accounted for by shares in the hands of 2 million people.

The structure of the Tata Group's ownership made it what Mr. Chugh called "a great example of innovation in an institution." All Tata Group companies were held by two "promoter" companies, Tata Sons, which has the group's main oper-ating companies under it, and Tata Industries, which promoted the group's entry into new businesses. Ownership of these two entities breaks down as follows:

• **Tata Sons** is held 66 percent by two public trusts, the Sir Dorabji Tata Trust and Sir Ratan Tata Trust; 18 percent by external shareholders; 13 percent by other Tata companies; and 3 percent by the Tata family.
• **Tata Industries** is held 29 percent by Tata Sons; 52 percent by other Tata companies; and 19 percent by the Jardine Madison Group.

The two public trusts provide endowments for the creation of such national institutions as the Indian Institute of Science, which dates to 1911; the Tata Institute of Social Sciences, 1936; Tata Memorial Hospital, known for the quality of its cancer research, 1941; the Tata Institute of Fundamental Research, 1945; and the National Center for the Performing Arts, 1966. In addition, the trusts funds companies committed to spending nearly $100 million per year on social welfare, sponsoring volunteer programs, and similar activities.

Tata's Innovativeness in the Auto Sector

Among the group's industrial innovations is Tata Motors, which has recently made the transition from a specialist in diesel-powered commercial vehicles to a successful automaker. When group Chairman Ratan Tata announced his desire a decade earlier to make passenger vehicles, "everyone laughed," Mr. Chugh recalled, "telling him: 'Very few companies can make cars in the same culture and mindset that makes buses and trucks.'" The Tata Group head nonetheless pledged that the firm would do it—and do it on its own—with the result that Tata Motors is currently either second or third on India's auto market with a 28 percent share. At the heart of this effort was the company's engineering research center, which either developed or adapted innumerable technologies for use in Tata's vehicles.

Profile of Tata Auto Components

A second innovation announced by Ratan Tata in defiance of skepticism from "bigwigs" is a $2,000 car, test models of which are already running. Tata Auto Components, which Mr. Chugh represents, is deeply involved in this endeavor, looking at component-level technology around the world. When TACO was formed in 1996, India did not have the cutting-edge technologies required; in fact, its component industry was so fragmented it could not provide products at anywhere near world-class quality. Now, however, TACO has 16 global partners, four engineering centers, and 16 plants; is focused on building exports; and has been referred to as the Delphi to Tata Motors' GM. He noted that TACO has a fiercely independent structure, being at times obliged to work harder than its competition to sell to Tata Motors because "there is always an internal rivalry." The secret of its success has been learning diverse technologies and offering complete program management capabilities.

To indicate the extent to which TACO has acquired knowledge and enhanced its research and development—as well as its "innovative spirit"—through joint ventures, Mr. Chugh named a few of the company's U.S. partners:

- Johnson Controls Inc. (JCI), for seating systems;
- Owens Corning, for sheet-molded composites;
- Visteon, for lighting and engine induction systems; and
- Hendrikson, for bus and commercial-vehicle suspension systems, in a venture that had yet to go into production.

He also offered a quick panorama of TACO's product areas: interior plastics, seating systems, exteriors and composites, wiring harnesses, telematics, sheet-metal assemblies, engine cooling, exhaust systems, mirrors, control cables, springs, braking systems, and, new in 2006, batteries, CV suspensions, lighting, and engine induction.

Leveraging Indian Talent to Serve Global Markets

A unique business model runs as a common thread through all of TACO's endeavors, he said, formulating it thus: "Not only use the engineering talent in India, but leverage the engineering talent in India for a global business market." The model's key success factor is engineering competence. TACO currently employs over 800 engineers—a figure expected to rise to 1,500 in 2007 and to 2,500 in 2008—spread over four facilities: its own TACO Engineering Center and three jointly held with partners, Tata Johnson Controls Engineering Center, TACO FAURECIA Engineering Center, and TACO Visteon Engineering Center. Noting that JCI, FAURECIA, and Visteon compete fiercely among one another in some sectors of the auto components business, Mr. Chugh said that TACO is able to manage relationships with all three successfully because of the core strength in engineering and product development that it brings to the table.

A Collaborative Model: Tata Johnson Controls

As an illustration of how the collaborative model has been applied, he chose the engineering division of Tata Johnson Controls (TJC), the first company in which it has been tried and the only one in which it was completely mature. A venture owned 50 percent each by TACO and JCI in keeping with Tata's "respect for partnerships" and preference for joint contribution over unilateral control, TJC has 400 employees worldwide and 2005 sales of $16 million in engineering services alone. In addition to running one of the largest dedicated automobile product design centers in India, the joint venture has an engineering center co-located with JCI in the United States that employs 55 engineers and is headed by Mr. Chugh himself, as well as some 150 engineers in Europe and 40 to 50 more in Japan. In all parts of the world, the design and development work done by the venture across the fields of automotive seating, interiors, and electronics is determined by JCI's own requirements.

The key to TJC's success lies in coordinating the efforts of engineers spread around the globe as they carry on work on a given product development program around the clock—something with which it had had great difficulty in the two years immediately following its establishment in 1995. To overcome this challenge, JCI adapted its business operating system, working with a TJC engineer on site to execute processes that ensured communication between India and other locations. Although this model is much harder to implement in design engineering than in information technology, it has nonetheless been extremely successful.

Anatomy of a U.S.–India R&D Cooperation

Mr. Chugh then described a cooperative R&D project in the safety and electronics field that TACO has entered into with a U.S. partner. This partner's current product/service model applies exclusively to high-end market needs in the United States and Europe—he characterized its segment as one that might include Mercedes or BMW—and most of its offerings need to be adapted considerably for regional functionality and regulatory requirements, particularly since the latter differs greatly between the United States and Europe. The project calls for joint development of a low-cost, global platform capable of being:

- **Manufactured in a low-cost country (LCC).** "Electronics products can be made in Thailand or China much more cost-effectively than even in India if it's a mass run," he said, adding that TACO would "definitely" consider production outside India.
- **Rolled out in most markets with minimal adaptation.** "On the hardware side it's going to be a modular product," he explained, "and on the software side there'll be nothing sitting on the hardware." Using the case of a telematic product as an illustration, he said that suites of services in different price ranges might be available for loading onto the product.

The design phase of this project had just been completed, and the partners expected the rollout to take place soon. To exemplify TACO's existing R&D work, he provided a few examples of air vents it had designed. Two patents—one on an air vent for MG Rover, the other on universal mechanisms for air vents—dated to October 2005, while a third patent, on an air-vent mechanism for a vehicle yet to be launched, the X1, had been filed in March 2006.

TACO's Opportunities and Aspirations

TACO sees its opportunities for research and development in interiors, electronics, safety, and emissions. As is the way of the automotive industry, the company takes its cues from its customers rather than doing original research. It inquires as to their problems and goes back to the lab in search of solutions, hitting (in the process) upon patentable discoveries or novel applications. TACO's aspirations, Mr. Chugh indicated, are global; the company believes it needs to be not only in China, India, and the ASEAN region but also in the markets of the United States and Western Europe, all of which are looking for low-cost country sourcing.

Although low-cost country sourcing might seem to be a "shoot-and-ship kind of idea," that is not the case, Mr. Chugh said, because at its core is engineering design. TACO is getting contracts from the U.S. Big Three automakers who say, "Here is a black box," which means: "I'll tell you only the space, and you design the part. Give me prototypes that are tested and validated, then manufacture it and send it to us." While the Chinese are much more cost-effective at doing just the shoot and ship—that is, providing a drawing and design—TACO's unique selling proposition is its ability to do everything in the same package. In fact, design capability is TACO's core strength. Its future will be devoted to building a deep engineering and R&D base that can enable it to develop technology and innovative solutions for its customers, which is its constant focus.

Policy Making for Economic Opening: A Personal Account

Having finished his presentation on Tata's activities, Mr. Chugh asked the audience's indulgence as he shared a personal experience from around 10 years earlier, when he worked for ICICI Bank, formerly ICICI Ltd. India. He had joined that financial development institution in 1975, when it was engaged in the first export development study, which was funded by the World Bank. ICICI did a second such study several years later, but even then the government's policy was not ready for it, and it was put on the shelf with its predecessor. The concept entered into play around 1986, when rumblings regarding a liberalization similar to that which was to begin in 1991 were heard in government circles. In 1990, another export development program was drafted, and that one was to run for about 10 years.

Mr. Chugh explained that he was bringing this to the attention of the audience as an interesting example; the World Bank had called it the most successful such program ever involving cooperation among multiple agencies. Its diverse stakeholders ranged from government financial institutions to research and development institutions and from industry to academics and other individuals. The Confederation of Indian Industry (CII) had been among the partners, and Dr. Mashelkar had been associated with the program as well.

Mutual Benefit as the Platform for Change

The ICICI initiative owed its success to several elements. Upon the joint initiative of the World Bank and the Government of India, a policy framework was created by the Bank itself. Incentives were set up for all involved: ICICI had no interest in assigning 10 people to an export development program, nor did the government care to put up money, unless there was to be some benefit. Since at that time India had a shortage of foreign exchange, the World Bank provided funding on the condition that the policy initiatives be put in place—which meant that the economy had to be opened up. Ataman Aksoy, a World Bank economist whom he remembered as having done a "fabulous job," sat with Mr. Chugh and, together, they looked at everything that might be done in pursuit of that goal.

Synchronizing Public and Private Incentives

When implementation began, however, there were innumerable glitches. The approach demanded was simultaneously top-down and bottom-up. As to the former, every project brought to the table required a government initiative at the policy level if change was to occur. Reserve Bank of India (RBI) guidelines might, for instance, prevent a foreign technician from being brought in for a particular R&D project. In response, a steering committee was created on which were represented major institutions: the secretary (commerce), the secretary (finance), RBI, the Ex-Im Bank, and ICICI. The World Bank was extremely helpful in putting pressure on the government.

The bottom-up element consisted of microlevel intervention on a project-by-project basis. As an illustration, nobody at the time wanted to look at ISO-9000[7]; it was hard to get even one company to send people to the lead assessor program being run by CII's Total Quality Management division. Mr. Chugh recalled doing something "that no banker does: double financing." CII was financed to set up the program, and companies were given 50 percent matching grants—"'gambling money,' as Dr. Mashelkar put it"—to send their employees. What this produced was not so much a trickle-down effect as a "blowing-up effect," in the sense that companies had become so enthusiastic that they were currently willing to pay for the program on their own.

This was not the only sort of intervention to occur. When Hindustan Motors, having built a huge factory in Hallol (since taken over by GM) announced its intention to set up a joint venture with a foreign partner to manufacture trucks, the International Finance Corporation in Washington put its foot down. Hindustan Motors was told it had to fix its Ambassador passenger car business, which was on the skids, before it could get involved in trucks.

Serving Innovation by Breaking the Rules

"We brought in consultants, we brought in experts from overseas, we broke almost every rule in the game," Mr. Chugh said. "And that is the kind of innovation with which I'm very proud to be associated, earlier with ICICI and now with Tata." For innovation, he maintained, was not only a matter of organizations, whether smaller or larger; it resided, as Dr. Mashelkar had said, in the leadership, or entrepreneurship, or technopreneurship. "Call it whatever you want," he concluded. "It's the people who make the difference."

Dr. Good, observing that in view of the hour the closing reception would have to double as the question period, expressed her regret that the government ministers attending the symposium had been unable to stay until its end. A pair

---

[7]ISO 9000 is a family of ISO (the International Organization for Standardization) standards for quality management systems.

of important insights had emerged from the presentations offered by the business people on the final panel: that partnerships will not work if they are one-sided but must be built on mutual advantage; and that when the innovation engine of private enterprise is turned loose, it gets ahead of policy relatively fast. Thanking the speakers for sharing extraordinarily interesting case studies, she turned the microphone over to Dr. Wessner for final comments.

Closing Remarks

Charles W. Wessner
National Research Council

Dr. Wessner seconded both Dr. Good's thanks to the panelists and her comments on the lessons their talks had provided, telling her: "You're very right that we need to try and get some of the public-policy people to listen a little more fully." He expressed his gratitude in addition for the contributions of a number of people and organizations to the day's success: the Confederation of Indian Industry, whose cosponsorship of the conference had made possible both its scale and its quality; the Embassy of India, and particularly Ambassador Sen; and the U.S. Department of State, many of whose offices had been of help. Finally, he singled out Dr. Sujai Shivakumar of the STEP Board's staff, praising him not only as a colleague but as a partner without whose efforts the event would not have been possible.

The U.S.–Indian Relationship: A Rare Vibrancy

To conclude the symposium, Dr. Wessner offered a summary of the numerous and, indeed, exciting perspectives that it had provided. A fundamental one—that there was enormous interest in the U.S.–Indian relationship—was plain from the attendance of 400 and the participation of four ministers of both cabinet and subcabinet rank. "I don't think we've ever seen this level of vibrancy on both sides," he said, speaking on behalf of those among the attendees with extensive policy experience. Moreover, the two countries' relationship on the plane of government had great depth and breadth. Accounts by Secretary Bodman, Dr. Marburger, Minister Sibal, and Deputy Chairman Ahluwalia of shared interests and objectives, as well as of the range of activities already under way, had been extremely encouraging.

Turning to the private sector, Dr. Wessner pointed to the growth and the vitality of the India–U.S. R&D relationship in both directions and again stated his concurrence with Dr. Good, this time with her observation that industry had a capacity to get ahead of policy makers very quickly. Worthy of emphasis was that the day's discussion had shown India's potential to be far beyond call centers and other low-cost services. India's diversity of very high-end, cutting-edge capabilities in different sectors—whether automotive, information technology, or pharmaceutical—was not often reflected in the press. Similarly, the area of cooperation among educational institutions in India and the United States was in need of further exploration, and this was an effort that the National Academies might well undertake.

Impressing on Americans the Need for Change

As a final point, Dr. Wessner signaled that while the two nations had real opportunities to capture, inherent in this quest was a challenge to change. Professor Dahlman and others had been right to emphasize that "things [were] going better" in India; still, according to Dr. Wessner—who apologized for the liberty he was taking with grammar—things needed to go "much more better."

A related point, and one that Americans had a difficult time understanding, was that it was incumbent upon the United States to change its institutions as well, since those that had succeeded in the post-War period would not necessarily be the ones to carry it through the 21st century. Indeed, the current success of the United States is based on its past investment, and since the 1960s, the country had cut its national investment in R&D roughly in half. "That doesn't seem to be investing for the future in the way we'd all like to see," he cautioned, repeating: "We, too, have our challenges, and we hope to address them."

III

RESEARCH PAPER

India's Knowledge Economy in the Global Context[1]

Carl J. Dahlman
Georgetown University

INTRODUCTION

The rise of India as an emerging economic power is increasingly in the global headlines. This is due in part to its large population and impressive growth rates, not just in the past three years, but the past decade and a half. However, it is also due to India's increasing scientific and technological capability.

This paper assesses India's knowledge economy in the global context. To put the analysis in context, the second section quickly summarizes some of the key global trends. The third provides an overview of the Indian economy and its recent economic performance. The fourth presents India's rising economic power and briefly summarizes some of its advantages and challenges. The fifth section benchmarks India's position in the global knowledge economy using a four-part framework that includes the economic and institutional regime, education and training, the information infrastructure and its use, and the innovation system. It summarizes some of the key challenges and policy issues in the first three of these. The innovation system is analyzed in more detail in the sixth section. That analysis includes a quick overview of the innovation system as well as some of the key issues that need to be addressed. The seventh section summarizes some of the key opportunities for greater U.S.–India collaboration. The final section provides a very brief summary and conclusions.

---

[1]This paper is based in part on Carl Dahlman and Anuja Utz, *India and the Knowledge Economy: Leveraging Strengths and Opportunities*, Washington D.C.: World Bank, 2005, as well as ongoing work the author is doing on the environment for innovation in India as part of a follow-up study by the World Bank and as part of new book he is writing on India and China.

KEY GLOBAL TRENDS

India's rise needs to be seen in the broader context of some of the broader global trends affecting growth and competitiveness.

One of these is the increased importance of knowledge. The world is in the midst of what could be considered a knowledge revolution. It is not that knowledge has not always been important for growth and competitiveness, but that there has been a speeding up in the rate of creation and dissemination of knowledge.

A second key trend is an increase in globalization. The share of goods and services that are traded as a percentage of global GDP has increased from 38 percent in 1990 to 48 percent in 2004. This is the result of greater trade liberalization worldwide. However, it is also the result of reductions in transportation and communications costs that result from rapid advances in technology.

A third and related trend is that knowledge markets have become global. Products and services are increasingly designed and developed for global markets in order to recoup the research and development (R&D) investments. In addition, R&D itself is becoming increasingly globalized. This is not just an increase in joint authorship of technical papers by teams from different countries, or joint patenting. An increasing amount of R&D is now being done by multinationals in countries other than their respective home countries, and not just among developed countries. India and China in particular are also benefiting from this trend as they are becoming hosts to many R&D centers set up by multinational companies, as well.

In addition, thanks to the reduction in communications costs, there is an increasing trend to source many knowledge-intensive services in lower-cost developing countries. This is part of what is driving global offshoring of knowledge-intensive services, such as back office functions, as well as engineering design, and even contract innovation services.[2]

The result of these trends is that innovation and high-level skills are becoming the most important determinants of competitiveness. Thus countries such as India need to develop more explicit strategies to take advantage of the rapid creation and dissemination of knowledge and to develop their own stronger innovation capabilities.

THE INDIAN ECONOMY

The Indian economy has had a very impressive performance (Table 1). Between 1990 and 2000, it grew at an average annual rate of 6.0 percent. Between

---

[2]See Thomas Friedman, *The World is Flat: A Brief History of the 21st Century*, New York: W. H. Freeman, 2005. for a good description of some of the main ICT trends that are making the world more integrated (flatter).

TABLE 1 Growth of output overall and by sector (average annual % growth)

	GDP		Agriculture		Industry		Manufacturing		Services	
	1990–2000	2000–2004	1990–2000	2000–2004	1990–2000	2000–2004	1990–2000	2000–2004	1990–2000	2000–2004
Low income	4.6	5.5	3.1	2.7	4.9	6	5.8	6.5	5.9	6.7
India	**6.0**	**6.2**	**3.0**	**2.0**	**6.3**	**6.2**	**7.0**	**6.5**	**8.0**	**8.2**
Low middle income	5.2	6.0	2.6	3.8	6.4	7.3	NA	NA	5.1	5.4
China	10.6	9.4	4.1	3.4	13.7	10.6	NA	NA	10.2	9.8
Upper middle income	2.1	2.7	0.3	2.2	1.5	2.5	4.5	2.1	2.8	2.7
High income	2.7	2.0	1.0	-1.3	1.9	0.3	NA	0.7	3.0	2.0
World	2.9	2.5	1.8	2.1	2.4	1.4	NA	1	3.1	2.3

SOURCE: World Bank, *World Development Indicators 2006*, Washington, D.C.: World Bank, 2006, Table 4.1.

2000 and 2004, it grew at an average rate of 6.2 percent. In the past three years, it has grown at slightly over 8 percent. The sector that has been growing the fastest has been services.

Compared to China, the structure of the economy has not changed as rapidly. Twenty-five years ago the per capita income of these two giant economies was very similar. However, China has had a much faster rate or growth for a longer period of time and more rapid structural change (Table 2). To some extent, India has not followed the traditional pattern of a large increase in the share of industrial value added and then a shift to services. There has been a faster and earlier shift to services, driven in part by a rapid growth of high-value knowledge-intensive services (such as information technology [IT], banking, consulting, and real estate), although they account for only a very small share of India's very large labor force.

Another difference between India and other developing countries is that it is much less integrated into the global system through trade (Table 3). The contrast with China is again very stark as the share of trade of goods and services in the Chinese economy is more than twice that of India.

INDIA AS A RISING ECONOMIC POWER

India is a rising economic power, but one that has not yet integrated very much with the global economy. It has many strengths, but it also will be facing many challenges in the increasingly globalized, competitive, and fast changing global economy.

Figure 1 presents the current and projected size through 2015 of the world's 15 largest economies in terms of purchasing power parity (PPP) comparisons.[3] Using PPP exchange rates, India already is the fourth largest economy in the word. Moreover, using average growth rates for the period 1991–2003 to project future size, India surpasses Japan by the end of next year to become the third largest economy in the world. During the period projected, China (currently, the second largest economy), will become the largest economy, surpassing the United States by about 2013. However, it should be emphasized that past performance is not necessarily a good predictor of future performance—just of potential, as future reality is usually different than projected trend. Nevertheless, this projection based on PPP exchange rates is helpful to emphasize that India has great potential, but also faces competition, particularly from China. It is therefore useful to quickly take stock of India's strengths and challenges.

---

[3]Rather than using nominal exchange rates, the figure uses purchasing power exchange rates. PPP rates provide a better measure for comparing the real levels of expenditure across countries. They are derived from price surveys across countries that compare what a given basket of goods would cost and those results to impute the exchange that should be used.

TABLE 2 Structure of output, 1990 vs. 2004

	Agriculture		Industry		Manufacturing		Services	
	1990	2004	1990	2004	1990	2004	1990	2004
Low income	32	23	26	28	15	15	42	49
India	**31**	**21**	**28**	**27**	**17**	**16**	**41**	**52**
Low middle income	19	12	39	41	27	NA	42	46
China	27	13	42	46	33	NA	31	41
Upper middle income	10	6	39	32	22	20	51	62
High income	3	2	33	26	22	18	65	72
World	6	4	33	28	22	18	61	68

SOURCE: World Bank, *World Development Indicators 2006*, Washington, D.C.: World Bank, 2006, Table 4.1.

India's key strengths are its large domestic market, its young and growing population, a strong private sector with experience in market institutions, and a well-developed legal and financial system. In addition, from the perspective of the knowledge economy, another source of strength is a large critical mass of highly trained English-speaking engineers, business people, scientists, and other professionals, who have been the dynamo behind the growth of the high-value service sector.

However, India is still a poor developing country. Its per capita income in 2004 was just $674 and with a billion people, it accounted for 17 percent of the world's population. Its share of global GDP is less than 2 percent (using nominal exchange rates), and just 1 percent of world trade. Moreover, 80 percent of its population lives on less that $2 a day, and 71 percent is rural, with about 60 percent of the total labor force still engaged in agriculture.

TABLE 3 Integration with global economy (% of GDP)

	Merchandise Trade		Trade in Services		FDI	
	1990	2004	1990	2004	1990	2004
Low income	24.1	37.8	6.5	9.4	0.4	1.4
India	**13.1**	**25**	**3.4**	**8.2**	**0.1**	**0.8**
Lower middle income	31.5	57.5	6.2	10.3	0.7	2.7
China	32.5	59.8	2.9	7.0	1.0	2.8
Upper middle income	38.3	67.0	8.1	10.2	1.0	2.8
High income	32.3	41.5	8.0	10.5	1.0	1.3
World	32.4	44.9	7.8	10.5	1.0	1.6

SOURCE: World Bank, *World Development Indicators 2006*, Washington, D.C.: World Bank, 2006, Table 6.0.

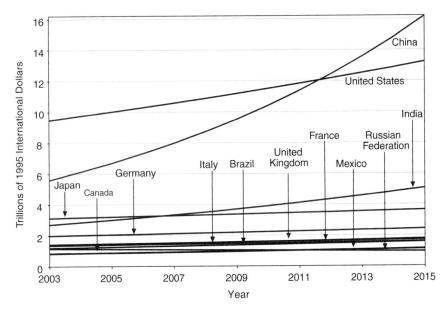

FIGURE 1 Current economic size and projection through 2015 for 15 largest economies. SOURCE: Author's projections based on data in the WDI database. World Bank, *World Development Indicators 2006*, Washington, D.C.: World Bank, 2006.

One of India's key challenges is its rapidly growing and young population. India's population is expected to continue to grow at a rate of 1.7 percent per year until 2020 and to overtake China as the most populous country in the world. Part of the challenge is that India's population has low average educational attainment. Years of school for the adult population averages less then 5 years, compared to nearly 8 years in China now, and 12 in developed countries. In addition, illiteracy is 52 percent among women and 27 percent among men.

Another challenges is poor infrastructure—power supply, roads, ports, and airports. This increases the cost of doing business. In addition, India is noted for an excessively bureaucratic and regulated environment which also increases the cost of doing business.

All these challenges constrain the ability of the Indian economy to react to changing opportunities. Low education reduces the flexibility to respond to new challenges. Poor infrastructure and high costs of doing business constrain domestic and foreign investment. The high costs of getting goods in or out of India also constrain India's ability to compete internationally and to attract export-oriented foreign investment except for business that can be done digitally rather than requiring physical shipments.

Figure 2 presents alternative projections of India's per-worker income to 2020. The projections assume that the growth of capital, labor, and education in

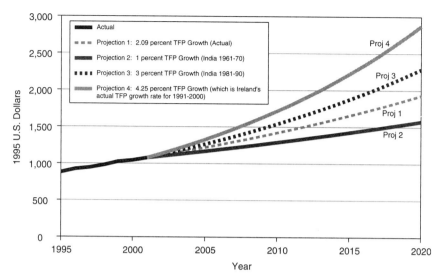

FIGURE 2 India's choice set in determining its future growth path: Real GDP Per Capita—Alternate projections, 2001-2020.
NOTE: The projections assume that capital, labor, and human capital (the educational complement to labor) grow at their 1991–2000 respective annual rates of growth. What is varied is the rate of total factor productivity growth. The TFP numbers are taken from the historical experience noted for each of the projections.
SOURCE: Carl Dahlman and Anuja Utz, *India and the Knowledge Economy: Leveraging Strengths and Opportunities*, Washington, D.C.: The World Bank, 2005.

India continue their trend lines. The only parameter that is changed is the rate of growth of total factor productivity (TFP)—the efficiency with which these basic factors are utilized.[4] The projections show that the real per-worker income in India could be between 46 to 167 percent higher in 2020 than in 2001, depending on how effectively knowledge is used. As noted, these projections are based on

---

[4]TFP is the residual to economic growth that remains after subtracting the rate of growth of capital, labor, and human capital. It is a broad indicator of the efficiency with which the other factors are used, and can be interpreted broadly as effectiveness in the use of knowledge, broadly defined to include technical, policy, and organizational knowledge. Four different sets of TFP growth are used for the projections. Projection 1 assumes a TFP growth rate of 2.09 percent, which was the average growth rate for India from 1991 to 2000. In this case real GDP per worker increased by 79 percent between 2001 and 2020. Projection 2 assumes a TFP growth rate of 1.05 percent, which was the rate of TFP growth for the India for 1961–1970. In this scenario real GDP per worker increases by 46 percent. Projection 3 assumes a TFP growth rate of 3 percent which was the average for India for 1981–1990. In this scenario real GDP per worker increases 112 percent. Projection 4 assumes a TFP growth rate of 4.25 percent, which was that achieved by Ireland for 1991–2000, a country which has been very successful at leveraging knowledge for its development. In this scenario real GDP per worker increases 167 percent.

the historical trends in the growth of inputs and of TFP. To a very large extent, these depend on policy measures that are under the control of India's policy makers, business, and the broader Indian society. The point of this projection is to emphasize that India's performance to a very large extent depends on its policy choices—what is holding India back is itself.

There is a tremendous window of opportunity for India to leverage its strengths to improve it competitiveness and increase the well-being of its population. However, it is important to seize these opportunities and to move quickly to action. The next section will examine India's position in the context of the global knowledge economy as a way to identify some of the key policy issues that need to be addressed to make India's recent rapid growth sustainable.

INDIA IN THE GLOBAL KNOWLEDGE ECONOMY

The World Bank Institute has developed a useful benchmarking tool that helps to rank countries in terms of their readiness to use knowledge for development.[5] The methodology consists of examining a country's rank ordering in four pillars based on a series of 20 indicators in each pillar. The four pillars are:

1. an economic and institutional regime that provides incentives for the efficient use of existing and new knowledge and the flourishing of entrepreneurship;
2. an educated and skilled population that can create, share, and use knowledge well;
3. a dynamic information infrastructure that can facilitate the effective communication, dissemination, and processing of information;
4. an efficient innovation system of firms, research centers, universities, consultants, and other organizations that can tap into the growing stock of global knowledge, assimilate and adapt it to local needs, and create new knowledge.

---

[5]See www.worldbank.org/kam. The knowledge assessment methodology (KAM) is designed to help countries understand their strengths and weaknesses in making the transition to the knowledge economy. It is thus useful in identifying the challenges and opportunities that a country faces, and where it may need to focus policy attention or future investments. In so doing, the KAM provides a preliminary knowledge economy assessment of a country, which can form the basis for more detailed sector-specific work. The KAM consists of a set of 80 structural and qualitative variables that serve as proxies for the four pillars that are critical to the development of a knowledge economy. The comparison is undertaken for a group of 128 countries which includes most of the developed Organisation for Economic Co-operation and Development economies and over 90 developing countries. The data used for this paper are from the 2006 version of the KAM. The basic scorecard for 14 variables is done for two points in time—1995 and the most current year for which data are available. See Figure A-1 in the Annex for the basic scorecard comparison of India, China, and the United States.

Broad Assessment of India's Position

A simple summary measure called the Knowledge Economy Index has been developed for quick comparative benchmarking. It is an amalgamated index consisting of the average ranking of three of the most indicative indicators for each of the four sectors.[6] This index is tracked over time. It permits the comparison of a country's current ranking to that in 1995. This is done in Figure 3 for India plus five other countries: Brazil, Russia, China, Korea, and Mexico[7] plus some other standard reference countries.

Figure 3 shows that India is placed roughly in the sixth decile of a rank-ordering distribution from the most advanced countries. It also shows that India's relative position has slipped relative to where it was in 1995. Figure 4 shows the contribution of each of the four pillars to India's relative ranking. India has improved its relative position on the innovation indicators and slightly on the information and communications technology (ICT) indicators. On the economic and institutional regime and education, it has slipped back. (See Annex Table A-1 for the ranking on each of the pillars.)[8]

The rest of this section summarizes very briefly some of the key issues in the economic and institutional regime, education and training, and information and communication technology. The following section looks at the issues in innovation in more detail.[9]

---

[6]The actual indices used are the following: For the economic and institutional regime: tariff and nontariff barriers as a proxy for the degree of competitive pressure; the rule of law and regulation as proxies for the effectiveness of government regulation. For education: literacy rates, secondary and tertiary enrollment rates. For ICT, fixed and mobile phone lines per 1,000 persons, computers per 1,000, and Internet users per 1,000. For the innovation pillar the three variables are: scientists and engineers in R&D, scientific and technical publications, and patents granted by the U.S. patent office. The latter is used because patent regimes differ so it was necessary to standardize for one regime. The United States was chosen because it was, until recently, the largest market. For the innovation pillar, only the methodology has two versions. In one, the three variables are scaled by population as are all the other variables in this summary indicator. The other uses the absolute numbers. This is the one used throughout this paper. The rationale is that for the innovation variables, absolute scale matters because knowledge is not consumed in its use. For more details on these and other variables, see the KAM Web site.

[7]This group, which is called the BRICKM countries, will be used throughout this paper as comparator countries for India. It has added Korea and Mexico but left out South Africa from the usual grouping of the so-called BRICS.

[8]A country can slip back even though it makes absolute progress in the specific area. This happens if the country's progress is less than that for the group as a whole. This is part of what has happened in the case of India for the education variables. There has been progress, but it has been less than that for the rest of the world. In some countries, sometimes there is an actual fall in the real values.

[9]For more details of the analysis, see Dahlman and Utz, 2005, op. cit., Chapter 2 on economic and institutional regime, Chapter 3 on education and skills, Chapter 4 on the innovation system, and Chapter 5 on the information infrastructure.

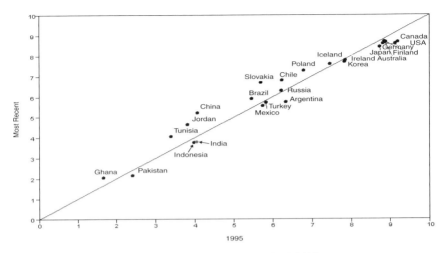

FIGURE 3 Changes to Knowledge Economy Index, 1995–2003.
NOTE: The horizontal axis represents the relative position of the country or a region in 1995. The vertical axis represents the position in the most recent year (generally 2000–2004). The graph is split by a 45 degree line. Those countries or regions that are plotted below the line indicate a regression in their performance between the two periods. The countries or regions that are marked above the line signify improvement between the two periods, while those countries that are plotted on the line indicate stagnation. The KAM methodology allows the user to check performance in the aggregate Knowledge Economy Index (KEI), as well as the individual pillars: Economic Incentive Regime, Education, Innovation, and ICT (Information Communications Technologies).
SOURCE: World Bank Institute, KAM 2006, <*http://www.worldbank.org/kam*>.

Key Issues in the Economic and Institutional Regime

The economic and institutional regime is an important aspect of a country's ability to take advantage of knowledge. It includes the overall regime of policies and institutions that give an economy the incentives to improve efficiency and the flexibility to redeploy capital and labor to their most productive use. It also includes the rule of law and government effectiveness. As was seen from the summary variables in the KAM basic scorecard, this is the second weakest of the four pillars of the knowledge economy in India, and one in which India has actually lost relative standing with respect to the rest of the world. Based on a more detailed analysis, including surveys of foreign and Indian businessmen, some of the key issues that have to be improved in the economic and institutional regime include:[10]

---

[10]See World Bank/International Finance Corporation, 2006, *Doing Business in 2006: Creating Jobs,* Washington, D.C.: International Bank for Reconstruction and Development, for how India compares to other countries on a large number of indicators of the domestic business environment.

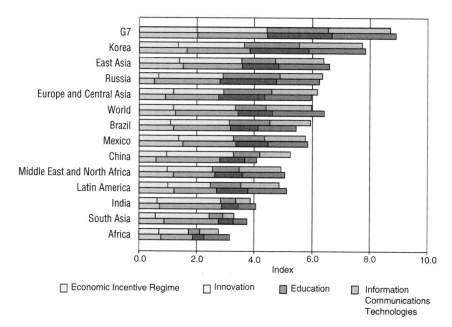

FIGURE 4 KEI: Major world regions and largest country in each, 1995 vs. most recent. NOTE: Each bar chart represents the most recent aggregate KEI score for a selected region or country, split into the four KE pillars. Each color band represents the relative weight of a particular pillar to the overall country's or region's knowledge readiness, measured by the KEI. The first line for each country is its position in the most recent year for which data are available (generally 2002–2005). The second line is for 1995. (See Annex Table A-1 for the actual ranking for each of the pillars. See Annex Figure A-1 for a comparison of the basic scorecard rankings for India with China and the United States.) SOURCE: World Bank Institute, KAM 2006, *<http://www.worldbank.org/kam>*.

- reducing the bureaucracy for the entry and exit of firms,
- updating physical infrastructure,
- easing restrictions on the hiring and firing of labor,
- reducing tariff and nontariff barriers to trade,
- encouraging foreign direct investment and increasing e-linkages with the rest of the economy,
- strengthening intellectual property rights and their enforcement, and
- improving e-governance and encouraging ICT use to increase government's transparency and accountability.

Key Issues in Education and Training

Educated and skilled persons underlie the ability of an economy to take advantage of knowledge and to create new knowledge to improve economic performance and welfare. Key elements of education and training for the knowledge economy include the level and quality of educational attainment as well as the relevance for the needs of a rapidly changing economy such as India. This is also a pillar in which India has slipped compared to its relative global ranking in 1995. Some of the key issues that India needs to address in education and training include:

- expanding quality basic and secondary education to empower India's rapidly growing young population;
- raising the quality and supply of higher education institutions, not just the Indian Institutes of Technology and the Indian Institutes of Management;
- embracing the contribution of private providers of education and training by relaxing bureaucratic hurdles and putting in place better accreditation systems;
- increasing university–industry partnerships to ensure consistency between education, research, and the needs of the economy;
- establishing partnerships between Indian and foreign universities to provide internationally recognized credentials;
- using ICT to meet the double goals of expanding access and improving the quality of education;
- investing in flexible, cost-effective job training programs that are able to adapt quickly to new and changing skill demands.

Key Issues in ICT

Advances in information processing, storage, and dissemination are making it possible to improve efficiency of virtually all information-intensive activities and to reduce transaction costs of many economic activities. Some of the key elements to make effective use of the potential of this new information infrastructure are the regulatory regime for the information and telecommunications industries and the skills to use the technologies, software, and applications. Some of the key issues that need to be improved in India include:

- boosting ICT penetration and reducing/rationalizing tariffs on hardware and software imports;
- massively enhancing ICT literacy and skills;
- increasing the use of ICT as a competitive tool to improve efficiency of production and marketing (supply chain management, logistics, etc.);
- moving up the value chain in IT by developing high-value products through R&D, improving the quality of products and services, marketing of products and services, and further positioning the "India" brand name;
- launching suitable incentives to promote IT applications for the domestic economy, including local language content and application;

  • strengthening partnerships between government agencies, research/
academic institutions, private companies, and nongovernmental organizations
(NGOs) to ramp up ICT infrastructure and applications;
  • developing/scaling up, through joint public–private partnerships, ICT applications, community radio, smart cards, Internet, satellite communications, etc.

STRENGTHENING INDIA'S INNOVATION SYSTEM

This section starts by placing India in the international context using the
KAM innovation pillar as well as other data. The next subsection develops a
brief framework for analyzing a developing country's innovation system. This
framework is then used to assess India's innovation system. The last section then
presents a matrix of key issues that need to be addressed to improve India's innovation system.

Broad Assessment of India's Position in Innovation

Figure 5 places India's innovation system in the global context using the
KAM innovation system pillars. This is based on one measure of R&D input
(scientists and engineers) and two measures of output (scientific and technical
publications, and patents in the United States). By this narrow measure linked
primarily to formal R&D, India is in the top 13th percentile of the global distribution of countries.[11] Furthermore, it has improved its position relative to the
rest of the world.

Clearly, because of India's large critical mass of scientists and engineers
engaged in R&D, India is a major player in global R&D. However, it is instruc-
tive to compare India's share of the world in scientists and engineers, scientific
and technical publications, and patents with its share of population and GDP
measured in nominal as well as PPP exchange rates (Figure 6). From this figure,
it can be seen that, as expected, India's share of scientists and engineers in R&D
is much lower than its share of population or GDP in PPP terms, although it is
slightly higher than its GDP share in nominal terms. Its share of scientific and
technical publications is smaller than its share of GDP in nominal terms. Its share
of all patents in the United States is extremely small (only 0.2 percent—too small
to be in the figure). One quick conclusion from this comparison is that India is
stronger in its basic scientific inputs that in its outputs of basic scientific and
technical knowledge, since its share of publications is smaller than its share of
personnel engaged in R&D. It is even weaker in turning that scientific output
into commercially relevant knowledge, as suggested by its much smaller share of

---

[11]However, its position would be much lower if measured relative to its population—see note to
Figure 5.

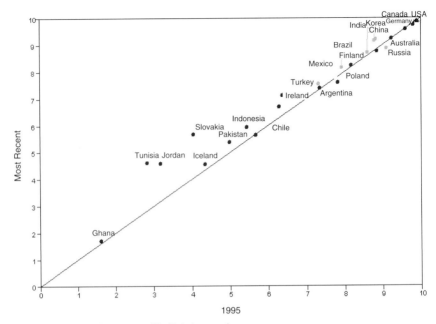

FIGURE 5 Global context of India's innovation system.
NOTE: This figure is based on the absolute size of India's innovative effort. If this were to be scaled by population (i.e., scientists and engineers in R&D per million population, scientific and technical publications per million population, patents in the United States per million population), India's relative position would fall to the 67th percentile of the country distribution.
SOURCE: World Bank Institute, KAM 2006, <*http://www.worldbank.org/kam*>.

patents in the United States. However, a developing-country's innovation system should be analyzed in a broader context, as developed below.

Components of a Developing County's Innovation System

A country's innovation system consists of the institutions and agents that create, adapt, acquire, disseminate, and use knowledge. It also includes the policies and instruments that affect the efficiency with which this is done. In developing countries, innovation should not be interpreted only as application of knowledge that is new at the level of the world frontier, but as product, process, organization, or business knowledge that is new to the local context. Therefore, in developing countries the innovation system should include not only domestic research and development and its commercialization and application. It should

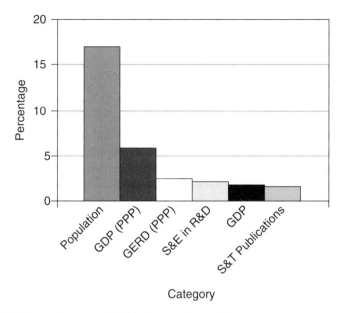

Category

FIGURE 6 Key indicators of India's share in the world.
SOURCE: Calculated from World Bank, *World Development Indicators 2006*, Washington, D.C.: World Bank, 2006.

also include the policies, institutions, mechanisms, and agents that affect the extent to which the country taps into and makes effective use of global knowledge that is new to the country.

The innovation system of a developing country such as India can be thought of as consisting of four parts. One is *formal R&D* that is carried out in India. This is the most visible and most easily measured. A second is the *informal innovation* in India. This may happen as the result of insights or experience by individuals or groups working in large of small enterprises or informal production. It can also be the result of decades of indigenous informal experimentation or accumulation of knowledge. This is not so visible and there is very little systematic quantification of this type of innovative effort. A third is *formal acquisition of foreign knowledge*. This includes the knowledge first brought in through direct foreign investment or technology transfer. The fourth is the *informal acquisition, adaptation, and use of knowledge* acquired through the import of capital goods, component products, and services that are new to the economy. It also includes knowledge obtained by copying, reverse engineering, or otherwise imitating what has already been done by others abroad. Other informal mechanisms include foreign study, travel, or work experience, as well as technical literature. Increasingly, it also includes all kinds of knowledge that can be acquired through the Internet including detailed manuals, designs, and data sets.[12]

Assessment of India's Innovation System

Table 4 compares some of the key indicators of India's broadly defined innovation system with that of the other BRICKM economies. China is the most relevant country for comparison because it is the closest in size and level of development. Figure 7 presents the main variables for India and China in graphical scorecard mode.[13]

Formal R&D

In India, the formal R&D effort is quite small. Total expenditures are only 0.8 percent of GDP and have been at that level for 15 years. The bulk of that effort (around 70–80 percent) is carried out by the public sector (federal and state), and most of that is mission-oriented R&D in defense, aerospace, and oceans. Only about 20 percent of that, or roughly 0.16 percent of GDP, is more applied work in agriculture, medicine, and industry.[14]

R&D spending by the private sector is only 16–20 percent of the total, or about 0.12 percent GDP. It is highly concentrated in a few large enterprises. The sectors that do the most R&D are pharmaceuticals, auto parts, electronics, and software.

A special feature is increasing R&D being done by multinational companies (MNCs) As of the end of 2004, there were nearly 200 R&D centers, including ABB, Astra Zeneca, Bell Labs Boeing, Bosch, Dell, Cummins, Dupont, Ericsson, Google, Honda, IBM, GE, GM Honda, Hyundai, Microsoft, Monsanto, Motorola, Nestle, Nokia, Oracle, Pfizer, Philips, Roche, Samsung, Sharp, Siemens, Unilever, and Whirlpool.[15] MNCs are attracted to set up R&D centers in India because of the lower salaries for Indian scientists and engineers, which are one-fourth to one-fifth that of comparable engineers in the United States.

---

[12]This framework was developed by the author for a forthcoming study on the environment for innovation in India being prepared by the World Bank. See the report for a more detailed application to India.

[13]In this graphical representation, the higher the index, the closer it is to the top of the global country distribution in that variable and the closer it is to the outside of the circle.

[14]Data from World Bank, *The Environment for Innovation in India*, South Asia Private Sector Development and Finance Unit, Washington, D.C.: World Bank, 2006.

[15]Data from Raja Mitra, "India's Potential as a Global R&D Power," in Magnus Karlsson (ed.), *The Internationalization of Corporate R&D*, Östersund: Swedish Institute for Growth Policy Studies, 2006.

TABLE 4 Innovation comparisons with BRICKMs

	Brazil	Russia	India	China	Korea	Mexico
Gross foreign investment as share of GDP (av. 1994–2003)	3.40	1.91	**0.70**	5.08	1.80	2.98
Royalty and license fee payments ($ million, 2004)	1,196.9	1,095.4	**420.8**	3,548.10	4,450.3	805.0
Royalty and license fee payments/million population (2004)	6.70	7.66	**0.40**	2.75	92.52	7.76
Royalty and license fee receipts (2004)	114.50	227.50	**25.20**	106.98	1,790.50	91.50
Royalty and license fee receipts/million population (2004)	0.64	1.59	**0.03**	0.08	37.22	0.88
Manufactured trade as % of GDP (2003)	15.10	17.83	**13.52**	51.32	48.65	45.99
High-technology exports as % of man. trade (2003)	11.96	18.86	**4.75**	27.103	32.15	21.34
Science and engineering enrollment ratio (% of tertiary students, 1998–2002)	NA	NA	**20.08**	NA	41.09	31.09
Science enrollment ratio (% of tertiary students, 1998–2003)	NA	NA	**15.11**	NA	10.25	12.52
Researchers in R&D (2003)	59,838	487,477	**117,528**	810,525	151,254	27,626
Researchers in R&D/million population (2002)	351.78	3,414.59	**119.66**	633.02	2,879.94	274.01
Total expenditures on R&D as % of GDP (2002)	1.04	1.24	**0.85**	1.23	2.91	0.43
Scientific and technical journal articles (2001)	7,205	15,846	**11,076**	20,978	11,037	3,209
Scientific and technical journal articles/million population (2001)	41.80	109.47	**10.73**	16.49	233.13	32.29
Patent applications granted by U.S. Patent and Trademark Office (2004)	161	173	**376**	597	4671	102
Patent applications granted by USPTO/million population (2004)	0.9	1.21	**0.35**	0.46	97.03	0.98

SOURCE: Compiled from World Bank Institute, KAM 2006, <http://www.worldbank.org/kam>.

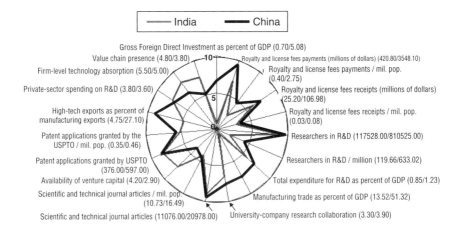

FIGURE 7 India-China comparison on selected indicators of innovation system. SOURCE: World Bank Institute, KAM 2006, <http://www.worldbank.org/kam>.

Informal Innovation

Informal innovation efforts are quite large. This consists not only of the experimentation and learning by doing that is done in the formal and informal sectors. There is very likely a grassroots innovation effort. Several NGOs have sprung up to support such grassroots innovation. They include Honeybee network, the Society for Research and Initiatives for Sustainable Development (SRISTI), and the Grassroots Innovation Augmentation Network (GIAN). In addition, the government has set up the National Innovation Foundation (NIF) to help document and finance grassroots innovations. The NIF has created a database of over 50,000 grassroots innovations. These consist of improvements in simple agricultural instruments, and agricultural techniques as well as indigenous knowledge. However, despite all these efforts, it has been difficult to develop appropriate funding and mechanisms to support the improvement, scale-up, and broad dissemination of grassroots innovations because of very high transaction costs and limited resources.[16]

Formal Acquisition of Foreign Knowledge

In India this has been small until relatively recently. For a long time, India has had a very strongly autarkic technology policy. There has been a gradual

---

[16]See a more detailed analysis in World Bank report, *The Environment for Innovation*, cited in footnote 14.

opening up of various parts of the economy to foreign investment. Now most sectors are open. The same is true for technology licensing, although there are still controls on the maximum royalty rates that can be charged. Until relatively recently, foreign investment into India was not allowed in many sectors, and was strictly regulated and kept to minority shares in joint ventures in others. There has been significant liberalization over the past 15 years, but India has not received as much foreign investment as the BRICKM countries. As can be seen from Table 4, gross foreign investment inflows as a share of GDP between 1994 and 2003 were the lowest among the six countries. Purchases of foreign technology have also been the lowest among the six countries, both in absolute terms and even more on a per capita basis. In addition, part of the reluctance of foreigners to invest in India, even after the sectors have been opened up, is the high degree of red tape, corruption, and bureaucracy as well as very poor physical infrastructure services. Some also worry about poor intellectual property rights enforcement.

Informal Acquisition of Foreign Knowledge

This is perhaps the most important source of domestic innovation in developing countries (except those that are very dependent on foreign investment such as Singapore and Hong Kong). As can also be seen in Table 4, India is again the least open economy of the six BRICKM countries as measured by degree of integration into the world economy through imports and exports of manufactured products. The share of manifested trade is only 13.5 percent of GDP in India compared to around 50 percent in China, Korea, and Mexico. Brazil and Russia are also less integrated with the global economy. However, these countries are outliers as the rest of the countries of the world are much more integrated into the global system (refer back to Table 3 for the share of merchandise trade and services in India compared to the average for other low-income countries, as well as lower and upper middle income countries, developed countries, and the world).

From Figure 7, comparing the key variables on the innovation system between India and China, it can be seen that China is ahead of India in virtually all the indicators, except the availability of venture capital, as well as some qualitative assessments on firm-level technology absorption and value chain reference where the persons surveyed have put India ahead.

However, in terms of the four-part framework laid out above, the following summary assessment can be made. It is hard to compare the domestic informal efforts, and so, that will be left aside. On acquiring knowledge from abroad informally, China is considerably ahead of India because it is much more integrated into the global system through trade and foreign education, and has a higher level of average educational attainment that facilitates the rapid assimilation of foreign knowledge. On acquiring foreign knowledge formally, China is also ahead because it has had a much more open policy for a longer period of time and has attracted much higher volumes of foreign investment as part of an explicit strategy to use foreign investment to produce new goods and services new to the Indian market, but also for exporting to the global market. Finally, in formal R&D effort, whereas China's spending as a share of GDP was comparable to India's in 1998, by 2005 it had been increased to 1.4 percent of GDP. China also plans to increase it further to 2.0 percent by 2010. In fact, in PPP terms, China in 2006 is probably already the second spender on R&D in the world, ahead of Japan and second only to the United States. Essentially, while China has been very effective at tapping global knowledge informally and informally leveraging these sources of innovation to improve its growth and welfare, it has now decided to do more to innovate on its own account, hence its major drive to increase formal R&D spending. Thus, it will be an even more formidable player on the global stage.

Key Areas for Strengthening India's Innovation System

Given the foregoing analysis, there is much that India needs to do to strengthen its innovation system. Time is of the essence given the trends and the increasing competitive demands of the global system, and the strategies of other countries—China in particular.

Table 5 summarizes in matrix form the main assessments made in the preceding section and proposes some areas for policy reform. The list is quite extensive. Furthermore, some of the proposed reforms get into areas where there may be considerable opposition and internal debate in India from various groups. Some of this is based on concerns about national sovereignty and ideology. Others are based on the concerns of groups with vested interests who want to maintain their position vis a vis new entrants, domestic as well as foreign. Thus, in a large complex democracy such as India, there will necessarily be a lot of debate. This process will take time. It is hoped that the analysis presented here can contribute to that debate and that concrete policies and investments will soon emerge.

OPPORTUNITIES FOR U.S.–INDIA COLLABORATION

There are many fertile areas for greater U.S.–India collaboration. These include trade, foreign investment, research, and education, and they are likely to increase as India advances in its reforms.

In trade, there is scope for increased exports and imports from each country to the other. Currently, trade levels are quite low, but the products and services produced by each country are very complementary so there is great potential to increase trade in both goods and services, particularly as India further liberalizes its trade regime.

There is also great scope for increased U.S. foreign investment in India as well as for more Indian investment in the United States. U.S. firms are already the largest investors in India, particularly in ICT service-related areas as well as in R&D centers. There is also much scope for increased strategic technological

TABLE 5 Summary of assessment and of areas in need of improvement

	Current Situation in India	Areas for Improvement
Creating knowledge domestically through formal R&D		
Government	Low public R&D expenditures relative to GDP	Increase public expenditures on R&D
	Low efficiency of public R&D expenditures	Improve the allocation and efficient use of public R&D
	Little transfer of knowledge created in public sector to productive sector	Strengthen institutions to commercialize knowledge Consider: • Bayh–Dole type legislation Strengthen: • technology transfer centers at universities and research institutes • science parks and business incubators
Indian firms	Still low but rising spending by productive firms	Encourage more R&D spending by productive firms through • promotion campaign on business advantages of R&D spending • more matching grants for R&D done by consortia • better fiscal incentives for more R&D spending
MNCs	Rapid increase in MNC R&D centers in India is creating shortages and increasing costs of scientific and technical personnel	Increase the supply of high-level scientific and engineering talent
Creating knowledge domestically through informal efforts		
Firms, formal and informal sector	Significant informal activity takes place, but there is little information or support	
Grassroots innovation and traditional knowledge, including NGOs and other networks	India has one of world's largest grassroots innovation systems, supported by Honeybee, GIAN, and SRISTI networks. However, there have been problems with scaling up and disseminating the innovations that come through this system	Strengthen institutional support through • training in technoentrepreneurship • laboratories for developing, piloting, and testing prototypes • funding for scale-up and dissemination

TABLE 5 continued

	Current Situation in India	Areas for Improvement
Acquiring knowledge from abroad through explicit contracts		
–FDI	FDI inflows into India are still relatively low in spite of increasing liberalization	Open sectors further to foreign investment.
	Foreigners are turned off by bureaucratic hurdles, red tape, corruption, poor infrastructure, and concerns about IPR enforcement	Improve the investment climate by reducing red tape and corruption and improving physical infrastructure and IPR enforcement
Strategic alliances	Beginning of some strategic alliances between foreign companies and domestic companies and research institutes	Increase strategic alliances by private and public sector. Requires more proactive marketing strategy
–Technology licensing	India has not made much use of foreign technology licensing	Increase formal technology licensing
Acquiring knowledge from abroad informally		
–Through Trade	India is still one of most closed economies of the world structurally (share of imports and exports in GDP) and in terms of tariff and nontariff barriers	Open economy further to trade by reducing tariff and nontariff barriers
–Through foreign education and training	Large numbers of Indian students go for tertiary education abroad. Many stay abroad. Some are starting to return	Develop good system to track students who go abroad for study. Launch public and private campaigns to attract them back by improving local salaries and working conditions
–Through more extensive use of Indian diaspora	There have been greater attempts to tap the Indian diaspora	Strengthen attempts to tap Indian diaspora
–Through technical literature	Access to foreign technical literature is limited by costs of books, technical publications, and databases	Exploit economies of scale in subscriptions through digital libraries and ICT network use

TABLE 5 continued

	Current Situation in India	Areas for Improvement
Through Internet	There is considerable access for more sophisticated users in large firms, universities, and research institutes, but this is constrained by low bandwidth even at high end, and there is still a low penetration rate of the Internet for the masses	Set up high-capacity research education network infrastructure Extend mass spread of Internet penetration by lowering costs, and set up multiple-use Internet kiosks and service centers

alliances between firms from the two countries. Some of the sectors in which there is strong potential for greater collaboration include pharmaceuticals, engineering goods, automobiles and auto parts, telecommunications equipment and services, and software.

There is also potential for greater collaboration between the United States and India in joint research on energy, environment, and space and in fact, several major agreements have recently been initiated between the two countries. Furthermore, given India's needs and experience and its large public research institute infrastructure, there is scope for joint work on major public good initiatives in health and preventive medicine as well as in agriculture and sustainable livelihoods.

In addition, there are many opportunities in higher education, including joint degrees, joint ventures, wholly owned subsidiaries or franchises. Furthermore, these are not just from the United States into India, but also from India to the United States. For example, NIT has set up many training facilities and developed specialized corporate training activities in the United States.

CONCLUSION

In sum, India has made great progress but faces daunting challenges. India has many strengths, particularly a young and growing population, experience and institutions of a market economy, a critical mass of entrepreneurs and highly skilled professionals, and a large public research infrastructure. It has the potential to leverage its strengths to improve its competitiveness and welfare. It faces many internal challenges as well as a much more demanding and competitive international environment.

This paper has presented a quick overview of the broad range of issues where India needs to deepen its economic reforms and make additional investments. It has assessed in a little more detail some of the key issues in its innovation system, and identified specific areas that need improvement.

There is also tremendous potential for increased U.S.–India cooperation across many areas. This conference is an opportunity to begin to develop this mutually beneficial cooperation. Hopefully this is just part of a series of events that will help to push the reforms and investment forward. Greater mutual understanding will spur greater public–public, public–private, and private–private cooperation, which will strengthen the mutually beneficial and strategic relationships between these two countries.

Annex

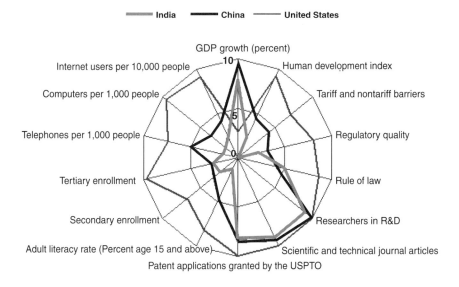

FIGURE A-1 Basic scorecard.
SOURCE: World Bank Institute, KAM 2006, <*http://www.worldbank.org/kam*>.

TABLE A-1 KAM ranking: How India compares with world regions and BRICKMs

Country	KEI	Economic Incentive Regime	Innovation	Education	ICT	KEI 1995	Economic Incentive Regime 1995	Innovation 1995	Education 1995	ICT 1995
G7	8.70	7.97	9.72	8.48	8.63	8.89	8.06	9.71	8.94	8.87
Korea	7.74	5.38	9.19	7.62	8.75	7.84	6.55	8.78	8.11	7.93
East Asia	6.38	5.54	8.68	4.62	6.68	6.59	6.08	8.22	5.05	7.01
Russia	6.33	2.68	8.91	7.85	5.88	6.22	2.05	9.09	7.78	5.95
Europe and Central Asia	6.17	4.77	6.95	6.67	6.27	5.96	3.66	7.32	6.45	6.41
World	5.99	4.77	8.60	4.26	6.33	6.40	5.04	8.67	4.74	7.14
Brazil	5.94	4.34	8.18	5.59	5.64	5.46	4.75	7.92	3.85	5.30
Mexico	5.74	5.43	7.59	4.37	5.58	5.82	6.07	7.31	4.40	5.52
China	5.24	3.84	9.24	3.60	4.30	4.07	2.32	8.82	3.48	1.68
Middle East and North Africa	4.93	3.91	6.24	3.71	5.84	5.05	4.76	5.67	3.83	5.93
Latin America	4.85	4.02	5.91	4.20	5.28	5.11	4.84	5.87	4.31	5.42
India	3.83	2.47	8.74	2.16	1.96	4.06	2.86	8.59	2.38	2.40
South Asia	3.30	2.27	7.46	1.88	1.58	3.74	3.49	7.56	2.03	1.88
Africa	2.72	2.78	4.05	1.51	2.55	3.14	2.99	4.41	1.61	3.56

SOURCE: World Bank Institute, KAM 2006, <http://www.worldbank.org/kam>.

IV

APPENDIXES

Appendix A

Biographies of Speakers[*]

MONTEK SINGH AHLUWALIA

Montek Singh Ahluwalia has been a key figure in India's economic reforms from the early 1980s onward. He is currently the deputy chairman of the Planning Commission for India, having been appointed to this post on June 16, 2004. He was previously the first director of the Independent Evaluation Office at the International Monetary Fund (IMF), a position he assumed on July 9, 2001. Prior to joining the IMF, Mr. Ahluwalia was a member of the Planning Commission in New Delhi as well as a member of the Economic Advisory Council to the prime minister. Before that, he served as finance secretary in the Ministry of Finance, secretary in the Department of Economic Affairs, commerce secretary, special secretary to the prime minister, and economic advisor to the Ministry of Finance. Between 1968 and 1979, he held various positions in the World Bank Research Department.

He earned his B.A. (Hon.) degree in New Delhi and his M.A. and M.Ph. degrees from the University of Oxford, where he was a Rhodes Scholar. His published works include papers in various professional journals and several contributions to books.

ROBERT ARMSTRONG

Robert Armstrong received his B.S. degrees in chemistry and biochemistry from the University of California at San Diego in 1979. He then moved to Colorado State University, where he received his Ph.D. degree in 1984 and

---

subsequently completed a postdoctoral fellowship at Harvard University. He joined the UCLA faculty in 1986 as an assistant professor in the Department of Chemistry and Biochemistry and was tenured in 1992. While at UCLA, Professor Armstrong's research efforts focused in the areas of synthesis and mechanism of action of bioactive compounds. Dr. Armstrong joined the senior research management team at Amgen, Inc. in 1996 and was responsible for developing Amgen's small-molecule drug discovery efforts. As of 1999, Dr. Armstrong has held the position of vice president, Discovery Chemistry Research and Technologies and Global External Research and Technologies at Eli Lilly and Co.

GEORGE ATKINSON

Dr. George H. Atkinson, named by then-Secretary of State Colin Powell to be science and technology adviser to the secretary (STAS) in September 2003, has continued to serve as STAS under Secretary of State Condoleezza Rice. The STAS is a principal interlocutor for science and technology with the U.S. Department of State. Dr. Atkinson joined the Department of State in August 2001, as the first American Institute of Physics Senior Fellow for Science, Technology, and Diplomacy. He continues efforts to strengthen the Department of State's scientific capacity by increasing the number of scientists in the department, introducing new anticipatory, proactive programs, and developing key domestic and international science and technology relationships.

Dr. Atkinson received his Ph.D. in chemistry from Indiana University in Bloomington. He was professor of chemistry at Syracuse University until 1983, when he joined the University of Arizona as professor of chemistry and optical sciences and head of the Chemistry Department. He remains a tenured professor at the University of Arizona.

Dr. Atkinson has more than 160 publications in refereed scientific journals and books, and has authored 66 U.S. and foreign patents. He also founded Innovative Lasers Corporation in 1992. His numerous honorary awards include the Senior Alexander von Humboldt Award (Germany), the Senior Fulbright Award (Germany), the Lady Davis Professorship (Israel), the SERC Award (Great Britain), the Distinguished Alumni Award for Outstanding Service from Indiana University, and the Chancellor's Distinguished Fellows Award from the University of California at Irvine. He has been a visiting professor at universities and research institutions in Japan, Great Britain, Germany, Israel, and France. In 1992, students selected him as the "outstanding teacher at the University of Arizona."

SAMUEL BODMAN

Samuel Wright Bodman was sworn in as the eleventh secretary of energy on February 1, 2005, after the U.S. Senate unanimously confirmed him on January

31, 2005. He leads the Department of Energy with a budget in excess of $23 billion and over 100,000 federal and contractor employees.

Previously, Secretary Bodman served as deputy secretary of the Treasury beginning in February 2004. He also served the Bush administration as the deputy secretary of the Department of Commerce beginning in 2001. A financier and executive by trade, with three decades of experience in the private sector, Secretary Bodman was well suited manage the day-to-day operations of both of these cabinet agencies.

Born in 1938 in Chicago, he graduated in 1961 with a B.S. in chemical engineering from Cornell University. In 1965, he completed his Sc.D. at Massachusetts Institute of Technology. For the next six years he served as an associate professor of chemical engineering at MIT and began his work in the financial sector as technical director of the American Research and Development Corporation, a pioneer venture capital firm. He and his colleagues provided financial and managerial support to scores of new business enterprises located throughout the United States.

From there, Secretary Bodman went to Fidelity Venture Associates, a division of the Fidelity Investments. In 1983, he was named president and chief operating officer of Fidelity Investments and a director of the Fidelity Group of Mutual Funds. In 1987, he joined Cabot Corporation, a Boston-based Fortune 300 company with global business activities in specialty chemicals and materials, where he served as chairman, CEO, and as a director. Over the years, he has been a director of many other publicly owned corporations.

Secretary Bodman has also been active in public service. He is a former sirector of MIT's School of Engineering Practice and a former member of the MIT Commission on Education. He also served as a member of the Executive and Investment committees at MIT, a member of the American Academy of Arts and Sciences, and a trustee of the Isabella Stewart Gardner Museum and the New England Aquarium.

RALPH CICERONE

Ralph J. Cicerone, president of the National Academy of Sciences, is an atmospheric scientist whose research in atmospheric chemistry and climate change has involved him in shaping science and environmental policy at the highest levels nationally and internationally.

His research was recognized on the citation for the 1995 Nobel Prize in chemistry awarded to University of California at Irvine colleague F. Sherwood Rowland. The Franklin Institute recognized his fundamental contributions to the understanding of greenhouse gases and ozone depletion by selecting Cicerone as the 1999 laureate for the Bower Award and Prize for Achievement in Science. One of the most prestigious American awards in science, the Bower also recognized his public policy leadership in protecting the global environment. In 2001, he led a National Academy of Sciences study of the current state of climate change and its impact on the environment and human health, requested by President Bush. The American Geophysical Union awarded him its 2002 Roger Revelle Medal for outstanding research contributions to the understanding of Earth's atmospheric processes, biogeochemical cycles, or other key elements of the climate system. In 2004, the World Cultural Council honored him with another of the scientific community's most distinguished awards, the Albert Einstein World Award in Science.

During his early career at the University of Michigan, Cicerone was a research scientist and held faculty positions in electrical and computer engineering. In 1978, he joined the Scripps Institution of Oceanography at the University of California at San Diego as a research chemist. From 1980 to 1989, he was a senior scientist and director of the atmospheric chemistry division at the National Center for Atmospheric Research in Boulder, Colorado. In 1989, he was appointed the Daniel G. Aldrich Professor of Earth System Science at the University of California at Irvine and chaired the Department of Earth System Science from 1989 to 1994. While serving as dean of physical sciences for the next four years, he brought outstanding faculty to the school and strengthened its curriculum and outreach programs. Prior to his election as Academy president, Cicerone was the chancellor of the University of California at Irvine from 1998 to 2005.

Cicerone is a member of the National Academy of Sciences, the American Academy of Arts and Sciences, and the American Philosophical Society. He has served as president of the American Geophysical Union, the world's largest society of earth scientists, and he received its James B. Macelwane Award in 1979 for outstanding contributions to geophysics. He has published about 100 refereed papers and 200 conference papers, and has presented invited testimony to the U.S. Senate and House of Representatives on a number of occasions.

Cicerone received his bachelor's degree in electrical engineering from the Massachusetts Institute of Technology where he was a varsity baseball player. Both his master's and doctoral degrees are from the University of Illinois in electrical engineering, with a minor in physics.

CARL DAHLMAN

Carl Dahlman is the Luce Professor of International Affairs and Information Technology at the Edmund A. Walsh School of Foreign Service at Georgetown University. He joined Georgetown in January 2005 after more than 25 years of distinguished service at the World Bank.

At Georgetown, Dahlman's research and teaching explore how rapid advances in science, technology, and information are affecting the growth prospects of nations and influencing trade, investment, innovation, education, and economic relations in an increasingly globalizing world.

At the World Bank, Dahlman served as senior advisor to the World Bank Institute. In this role he managed the Knowledge for Development (K4D) program, an initiative providing training on the strategic use of knowledge for economic and social development to business leaders and policy makers in developing countries. Prior to developing the K4D program, Dahlman served as staff director of the 1998–1999 World Development Report, Knowledge for Development. In addition, he was the bank's resident representative and financial sector leader in Mexico from 1994 to 1997, years during which the country coped with one of the biggest financial crises in its history. Before his position in Mexico, Dahlman led divisions in the bank's Private Sector Development, and Industry and Energy Departments. He has also conducted extensive analytical work in major developing countries including Argentina, Brazil, Chile, Mexico, Russia, Turkey, India, Pakistan, China, Korea, Malaysia, Philippines, Thailand, and Vietnam.

Dahlman's publications include: *India and the Knowledge Economy: Leveraging Strengths and Opportunities* (2005), *China and the Knowledge Economy: Seizing the 21st Century (2001)*, and *Korea and the Knowledge-Based Economy: Making the Transition* (2000). He is currently finishing a knowledge economy study on Mexico, working on a book on the challenge of the knowledge economy for education and training in China, and collaborating with research teams in Finland, Japan, and Korea to produce books on each country's innovation and development strategies.

Dahlman earned a B.A. magna cum laude in international relations from Princeton University and a Ph.D. in economics from Yale University. He has also taught courses at Columbia University's School of International and Public Affairs.

PAULA J. DOBRIANSKY

Paula J. Dobriansky was nominated by President Bush on March 12, 2001, unanimously confirmed by the Senate on April 26, and on May 1, sworn in as undersecretary of state for global affairs. On July 29, 2005, she became undersecretary of state for democracy and global affairs. In this capacity, she is responsible for a broad range of foreign policy issues, including democracy, human rights, labor, refugee and humanitarian relief matters, and environmental/science issues. She has also been designated as the special coordinator for Tibetan issues.

Prior to her appointment, Dr. Dobriansky served as senior vice president and director of the Washington Office of the Council on Foreign Relations. She was responsible for managing the Council's office and operations in Washington, D.C., and for leading council meetings, study groups, and seminars that served over 1,000 area members. She was also the council's first George F. Kennan Senior Fellow for Russian and Eurasian Studies.

Previously, Dr. Dobriansky served as senior international affairs and trade advisor at the law firm of Hunton & Williams, and also as cochair of the International TV Council at the Corporation for Public Broadcasting. Her government appointments include associate director for policy and programs at the United States Information Agency, deputy assistant secretary of state for human rights and humanitarian affairs, deputy head of the U.S. delegation to the 1990 Copenhagen Conference on Security and Cooperation in Europe, advisor to the U.S. delegation to the 1985 U.N. Decade for Women Conference in Nairobi, Kenya, and director of European and Soviet affairs at the National Security Council, the White House.

Dr. Dobriansky received a B.S.F.S. summa cum laude in international politics from Georgetown University School of Foreign Service and an M.A. and a Ph.D. in Soviet political/military affairs from Harvard University. She is a Fulbright-Hays Scholar, Ford and Rotary Foundation Fellow, a member of Phi Beta Kappa, and a recipient of various honors, including Georgetown University's Annual Alumni Achievement Award, the State Department's Superior Honor Award, Dialogue on Diversity's International Award 2001, National Endowment for Democracy (NED) Democracy Service Medal, Poland's Highest Medal of Merit, Grand Cross of Commander of the Order of the Lithuanian Grand Duke Gediminas, honorary doctorate of humane letters from Fairleigh Dickinson University, Westminster College, Roger Williams University, and an honorary doctorate of laws from Flagler College.

Dr. Dobriansky has served on various boards, including the Western NIS Enterprise Fund, National Endowment for Democracy (vice chairman), Freedom House, American Council of Young Political Leaders, the American Bar Association Central/East European Law Initiative, and the U.S. Advisory Commission on Public Diplomacy. She has a working knowledge of French, Russian, Italian, and Dutch.

Dr. Dobriansky has lectured and published articles, book chapters, and op-ed pieces on foreign affairs-related topics, ranging from U.S. human rights policy to East European foreign and defense policies, public diplomacy, democracy promotion strategies, Russia, and the Ukraine. For three years, she hosted *Freedom's Challenge* and cohosted *Worldwise*, the international affairs programs on National Empowerment Television. Additionally, she has appeared on ABC, NBC, CBS, CNN Headline News, CNN & Company, Fox Morning News, John McLaughlin's One-on-One, The McLaughlin Group, C-SPAN, MSNBC, PBS, National Public Radio, and has testified often before the Senate Foreign Relations and House International Relations Committees.

MARY GOOD

Mary L. Good is the Donaghey University Professor at the University of Arkansas at Little Rock, and serves as dean for the College of Information Science and Systems Engineering. She is managing member for the Fund for Arkansas' Future, LLC. (an investment fund for start-up and early-stage companies), past president of the American Association for the Advance of Science, past president of the American Chemical Society, and an elected member of the National Academy of Engineering. She presently serves on the boards of BiogenIdec, Inc. and Acxiom, Inc.

Previously, she served a four-year term as the undersecretary for technology for the Technology Administration in the Department of Commerce, a presidentially appointed, Senate-confirmed, position. In addition, she chaired the National Science and Technology Council's (NSTC) Committee on Technological Innovation, and served on the NSTC Committee on National Security. Previously, she has served as the senior vice president for technology for Allied Signal and as the Boyd Professor of Chemistry and Materials Science at Louisiana State University.

She was appointed to the National Science Board by President Carter in 1980 and by President Reagan in 1986. She was the chair of that board from 1988 to 1991, when she received an appointment by President Bush to be a member of the President's Council of Advisors on Science and Technology.

She has received many awards, including the National Science Foundation's Distinguished Public Service Award, the American Institute of Chemists' Gold Medal, the Priestly Medal from the American Chemical Society, and the Vannevar Bush Award from the National Science Board, among others.

Good received her bachelor's degree in chemistry from the University of Central Arkansas and her M.S. and Ph.D. degrees in inorganic chemistry from the University of Arkansas at Fayetteville.

GOPAL GOPALAKRISHNAN

Dr. P. (Gopal) Gopalakrishnan is the director of the IBM India Research Laboratory, which is part of IBM Research, widely recognized as the world's leading IT research organization. Gopal has over 18 years of experience in technology and research, and now leads a team of researchers in developing innovative technologies for IBM products and services and in addressing the unique issues faced by clients in the region. The India Research Lab has projects that span several important areas of technology: software, systems, and services. The newly created Services Innovation and Research Center of the India Research Lab is colocated with IBM's global services teams in Bangalore and focuses on research and development of technologies to increase the competitiveness of IBM's services organizations.

Prior to this position, Gopal led IBM's research strategy in pervasive computing and managed the pervasive infrastructures department at the IBM Thomas J. Watson Research Center in Yorktown Heights, New York. In this role, he managed the development of advanced technologies in infrastructure middleware, device components, and prototype solutions for pervasive computing and set the directions of researchers across the worldwide labs of IBM in this area of research. Gopal joined the IBM T. J. Watson Research Center in 1986 after earning a Ph.D. in computer science from the University of Maryland. He made significant contributions to the field of speech recognition and conversational interfaces while part of a long-running project at IBM Research. Over the course of his research career, he has worked in several disciplines, including parallel processing, speech recognition, conversational systems, and mobile computing. He is the coinventor on 18 patents and has authored many technical publications in peer-reviewed journals and conferences.

KENNETH G. HERD

Dr. Kenneth G. Herd was named global technology leader for the Material Systems Technologies at GE Global Research in February 2006. Dr. Herd's organization develops breakthrough material systems and material processes for a range of GE products, including composites for GE-Aviation and GE-Wind Energy, optical films and optical storage media for GE-Plastics, x-ray sources for GE-Healthcare and GE-Security, and material inspection and modeling for GE-Inspection Technologies. Dr. Herd leads a multidisciplinary, global organization of about 300 engineers and scientists, including 10 labs at GE Global Research in Niskayuna, New York, six labs at the Jack F. Welch Technology Center in Bangalore, India, and two labs at GE's China Technology Center in Shanghai.

He began his General Electric career in 1983, working on the development of GE's first magnetic resonance imaging systems for GE-Healthcare, moving to GE-Energy in 1986 to develop high-performance generators. In 1988, he joined the Electro-Mechanical Systems Lab at GE Global Research to develop superconducting materials and devices.

In 1998, Dr. Herd was named manager of the Measurement Systems Lab, leading the development of ultrasound, x-ray, and infrared imaging systems for industrial applications. He assumed the position of global technology leader for Inspection and Manufacturing Technologies in 2001, leading 10 labs in the development of a broad range of material process technologies.

Dr. Herd earned his bachelor's and master's degrees in mechanical engineering in 1981 and 1983 from the University of Massachusetts in Amherst, and his doctorate in mechanical engineering from Rensselaer Polytechnic Institute in 1991. Dr. Herd holds 44 U.S. patents and has published 26 papers.

P. V. INDIRESAN

Professor Indiresan was educated in Presidency College, Madras, Indian Institute of Science, Bangalore, and at the University of Birmingham, UK. He has taught for 40 years, starting his career at the University of Roorkee and then shifting to IIT Delhi. He has been a visiting professor at the Imperial College,

London, and a fellow at the Wissenschaftskolleg at Berlin. For a term, he was director, IIT Madras.

At IIT Delhi, Professor Indiresan founded the School of Radar Studies, now renamed as the Centre for Applied Research in Electronics. He also served there as a dean for examinations and as dean of undergraduate studies.

Professor Indiresan has been president of the Institution of Electronics and Telecommunication Engineers, and of the Indian National Academy of Engineering.

He is a fellow of the Indian National Academy of Engineering, a distinguished fellow of Institution of Electronics and Telecommunication Engineers, and honorary fellow of the Indian Railway Society of Signal and Telecommunication Engineers.

Professor Indiresan was twice awarded the highest prize of the Inventions Promotion Board of the government of India. He has also been conferred the Padma Bhushan by the president of India. He has the rare honor of being made an honorary member by the Institute of Electrical and Electronics Engineers, USA. His students have built a hostel in his name.

He is currently a member of the State Planning Commission, Delhi, a director of the Indo–Sri Lanka Foundation set up jointly by the two governments, and chairman, Netaji Institute of Science and Technology, an autonomous institution of the government of Delhi.

Professor Indiresan has written several hundred articles on societal and technical issues in Indian journals. He prepared the final report "Driving Forces and Impedances" of the Vision 2020 series compiled under the direction of President Kalam. He has a biweekly column in the *Hindu Businessline* and has written two books—*Managing Development: Geographical Socialism, Decentralisation and Urban Replication,* and *Vision 2020: What India Can Be, and How to Make That Happen*—as well as chapters in over 20 books.

SURINDER KAPUR

Dr. Surinder Kapur, chairman, National Mission on Manufacturing Innovation, and founder chairman and managing director, Sona Group, is representing the SME sector on the National Manufacturing Competitiveness Council. An engineer by qualification, Dr. Kapur has led quality and innovation in Indian industry through the various committees and initiatives of CII. He is leading CII's initiative of SME Clusters under its TQM program as well as advanced clusters on breakthrough management (innovation).

Under his chairmanship of the Mission on Manufacturing Innovativeness, CII will create 100 leader companies in Indian industry with global processes on innovation and product development over three years by setting up an Innovation Center of Excellence. Dr. Kapur also chairs the Automotive Component Manufacturers Association (ACMA) of India Centre for Technology. As a member of the National Automotive Testing and R&D Infrastructure Project, an initiative by the government of India to address the critical gap in testing and R&D infrastructure in the country, he is leading the Automotive Component Manufacturers' efforts to support government-funded R&D activities through the Technology Information, Forecasting, and Assessment Council. As a member of the Technical Advisory Committee on Automotives, DGTD and the Development Council for Automotives & Allied Industries GOI, Dr. Kapur has made a number of policy recommendations.

Dr. Kapur and his company have been the recipients of a number of awards and recognitions in the areas of quality, manufacturing excellence, and leadership. His pursuit of quality led the Sona Group to winning the Deming Award from the Japanese Union of Scientific Engineers and the Frost & Sullivan Corporate Gold Award for Excellence in Manufacturing. Dr. Kapur has been instrumental in Sona Koyo receiving the 'Technology Award' from AMCA in 2004 due to six patents filed by Sona Koyo.

Dr. Kapur has endeavored to integrate quality into Indian industry as the chair of the CII National Committee on Quality & Training Services, the TPM Club of India and AMCA.

He studied mechanical engineering at the University of Michigan and was the vice chairman and managing director of Bharat Gears from 1972 to 1990. He established Sona Koyo in 1990 and has since been leading his company from one success to another.

JOHN MARBURGER

Dr. John H. Marburger, III, science adviser to the president and director of the Office of Science and Technology Policy, was born on Staten Island, New York, grew up in Maryland near Washington, D.C., and attended Princeton University (B.A. in physics, 1962) and Stanford University (Ph.D. in applied physics, 1967). Before his appointment in the Executive Office of the President, he served as director of Brookhaven National Laboratory from 1998, and as the third president of the State University of New York at Stony Brook (1980–1994). He came to Long Island in 1980 from the University of Southern California where he had been a professor of physics and electrical engineering, serving as Physics Department chairman and dean of the College of Letters, Arts and Sciences in the 1970s. In the fall of 1994, he returned to the faculty at Stony Brook, teaching and doing research in optical science as a university professor. Three years later, he became president of Brookhaven Science Associates, a partnership between the university and Battelle Memorial Institute that competed for and won the contract to operate Brookhaven National Laboratory.

While at the University of Southern California, Marburger contributed to the rapidly growing field of nonlinear optics, a subject created by the invention of the laser in 1960. He developed theory for various laser phenomena and was a cofounder of the University of Southern California's Center for Laser Studies. His teaching activities included "Frontiers of Electronics," a series of educational programs on CBS television.

Marburger's presidency at Stony Brook coincided with the opening and growth of University Hospital and the development of the biological sciences as a major strength of the university. During the 1980s, federally sponsored scientific research at Stony Brook grew to exceed that of any other public university in the northeastern United States.

During his presidency, Marburger served on numerous boards and committees, including chairmanship of the governor's commission on the Shoreham Nuclear Power facility, and chairmanship of the 80-campus Universities Research Association, which operates Fermi National Accelerator Laboratory near Chicago. He served as a trustee of Princeton University and many other organizations. He also chaired the highly successful 1991–1992 Long Island United Way campaign.

While on leave from Stony Brook, Marburger carried out the mandates of the Department of Energy to improve management practice at Brookhaven National Laboratory. His company, Brookhaven Science Associates, continued to produce excellent science at the lab while achieving ISO14001 certification of the lab's environmental management system, and winning back the confidence and support of the community.

R. A. MASHELKAR

Dr. R. A. Mashelkar is presently the director general of the Council of Scientific and Industrial Research (CSIR), the largest chain of publicly funded industrial research and development institutions in the world, with 38 laboratories and about 20,000 employees.

Dr. Mashelkar is also the president of the Indian National Science Academy. He is only the third Indian engineer to have been elected as a Fellow of the Royal Society, London, in the twentieth century. He was elected foreign associate of U.S. National Academy of Sciences in 2005, only the eighth Indian since 1863 to be so elected. He was elected foreign fellow of U.S. National Academy of Engineering (2003), fellow of the Royal Academy of Engineering, UK (1996), and fellow of World Academy of Art & Science, USA (2000). Twenty universities, which include the Universities of London, Salford, Pretoria, Wisconsin, and Delhi, have honored him with doctorates.

In August 1997, Business India named Dr. Mashelkar as being among the 50 pathbreakers in the postindependent India. In 1998, Dr. Mashelkar won the JRD Tata Corporate Leadership Award, the first scientist to win it. In June 1999, *Business India* did a cover story on Dr. Mashelkar as *"CEO OF CSIR Inc.,"* a dream that he himself had articulated when he took over as director general of CSIR in July 1995. On November 6, 2005, he received the Business Week (USA) award of Stars of Asia at the hands of George Bush (Sr.), former President of United States.

When Dr. Mashelkar took over as the director general of CSIR, he enunciated *"CSIR 2001: Vision & Strategy."* This was a bold attempt to draw out a corporate-like R&D and business plan for a publicly funded R&D institution. This initiative has transformed CSIR into a user-focused, performance-driven, and accountable organization. This process of transformation has been recently heralded as one of the 10 most significant achievements of Indian Science and Technology in the twentieth century.

Dr. Mashelkar has been propagating a culture of innovation and balanced intellectual property rights regime for over a decade. It was through his sustained and visionary campaign that growing awareness of intellectual property rights (IPR) has dawned on Indian academics, researchers, and corporations. He spearheaded the successful challenge to the U.S. patent on the use of turmeric for wound healing and also the patent on Basmati rice. These landmark cases have set up new paradigms in the protection of India's traditional knowledge base, besides leading to the setting up of India's first Traditional Knowledge Digital Library. In turn, at an international level, this has led to the initiation of the change of the International Patent Classification System to give traditional knowledge its rightful place. As chairman of the Standing Committee on Information Technology of World Intellectual Property Organization, as a member of the International Intellectual Property Rights Commission of the UK government, and as vice chairman on the Commission in Intellectual Property Rights, Innovation and Public Health set up by the World Health Organization, he brought new perspectives to the issue of IPR and the developing world concerns.

In the postliberalized India, Dr. Mashelkar has played a critical role in shaping India's science and technology policies. He was a member of the Scientific Advisory Council to the Prime minister and also of the Scientific Advisory Committee to the cabinet set up by successive governments. He has chaired 10 high-powered committees set up to look into diverse issues of higher education, national auto fuel policy, overhauling the Indian drug regulatory system, dealing with the menace of spurious drugs, reforming the Indian agricultural research system, etc. He has been a much sought after consultant for restructuring the publicly funded R&D institutions around the world; his contributions in South Africa, Indonesia, and Croatia have been particularly notable.

Dr. Mashelkar has won over 40 awards and medals, which include the S. S. Bhatnagar Prize (1982), Pandit Jawaharlal Nehru Technology Award (1991), G. D. Birla Scientific Research Award (1993), Material Scientist of the Year Award (2000), IMC Juran Quality Medal (2002), HRD Excellence Award (2002), Lal Bhadur Shastri National Award for Excellence in Public Administration and Management Sciences (2002), World Federation of Engineering Organizations (WFEO) Medal of Engineering Excellence (2003) by WFEO Paris, Lifetime

Achievement Award (2004) by the Indian Science Congress, the Science Medal by the Third World Academy of Sciences.

The President of India honored Dr. Mashelkar with Padmashri (1991) and with Padmabhushan (2000), which are two of the highest civilian honors, in recognition of his contribution to nation building.

DAVID McCORMICK

David H. McCormick is the undersecretary of commerce for industry and security. Nominated by President Bush, he was confirmed by the U.S. Senate on October 7, 2005, and was sworn into office by Secretary of Commerce Carlos M. Gutierrez.

As undersecretary, Mr. McCormick leads the Bureau of Industry and Security, which advances U.S. national security, foreign policy, and economic objectives by ensuring an effective export control and treaty compliance system and promoting continued U.S. strategic technology leadership.

Prior to his service as undersecretary, Mr. McCormick was the president of Ariba, Inc., and had served previously as the president and CEO of FreeMarkets, Inc., both publicly traded software and services companies. Before joining FreeMarkets, Mr. McCormick was a consultant for McKinsey & Company.

Born in Pennsylvania, Mr. McCormick served as an officer in the U.S. Army and is a veteran of the first Gulf War. Mr. McCormick earned his bachelor's degree from the U.S. Military Academy at West Point and holds a master's degree and Ph.D. from the Woodrow Wilson School of Public and International Affairs at Princeton University.

PRAFUL PATEL

Praful Patel is regional vice president, South Asia Region, at the World Bank. Mr. Patel specializes in the development field, with an emphasis on lesser developed countries. Specific areas include infrastructure, poverty programs (MGD-related sectors), institution and capacity building, and multisectoral project design and packaging. Examples include the Chad Cameroon Pipeline, Zambia Copper Sector, and Africa Capacity Building.

Praful Patel, a Ugandan national, assumed his current position in July 2003. He oversees the Bank's Operations in Afghanistan, Bangladesh, Bhutan, India, Maldives, Nepal, Pakistan, and Sri Lanka. He joined the bank in January 1974 as part of the Young Professionals Program. Upon graduation from this program, he was appointed urban and regional planner in the Transport and Urban Projects Department, where he was promoted to deputy division chief in December 1979. In October 1984, he was promoted to the position of program coordinator in the Office of the Regional Vice President, Europe, Middle East, & North Africa

(MENA). In 1987, he was appointed division chief of the Infrastructure Operations Division in CD2 of the Asia Region. In February 1991, he was appointed country operations division chief in the Southern Africa Department. In May 1996, he became director of Finance, Private Sector and Infrastructure, Africa Region. From this position, he was promoted to his current position as regional vice president, South Asia.

Prior to joining the bank, Mr. Patel worked in Kenya in private practice and for the Housing Research and Development Unit at the University of Nairobi, and as Instructor at the Massachusetts Institute of Technology.

Mr. Patel's academic qualifications include a bachelor's degree in architecture (1st Class Honors) University of Nairobi, with a final-year program at Royal Academy of Fine Arts, Copenhagen (1971); M.A.A.S. (thesis on urban settlement design in developing countries) from MIT (1973), and General Manager Program at Harvard Business School (1996).

SWATI PIRAMAL

Dr. Swati A. Piramal is director, Strategic Alliances & Communications of Nicholas Piramal India Limited. Her current responsibilities include R&D, information technology, medical services, and knowledge management for the Healthcare Group of Piramal Enterprises.

A medical doctor (M.B.B.S.) from the University of Bombay, Dr. Piramal graduated with a master's degree from Harvard School of Public Health, where she had the unique honor of being selected commencement speaker at the 1992 graduation ceremony.

Dr. Piramal's special research interests include herbal, clinical discovery, and nutrition research in pharmaceuticals, and the use of management techniques such as information technology and communication to improve access and lower health care costs to meet the needs of the underprivileged children. Her specific research interests focus on malaria, tuberculosis, AIDS and diabetes.

Under her leadership, Nicholas Piramal has made significant progress in discovery research and patenting of new chemical entities, chemical process development for new drug delivery systems, clinical research for planning clinical trials, for new drug protocols and pharmacokinetics labs, herbal research for DNA fingerprinting and standardization of Ayurveda, the setting up of a business R&D program in the company, and contract research and technology partnerships with some of the leading companies in the world.

Dr. Piramal was part of the management team at Nicholas Piramal that acquired the Hoechst Marion Roussel Research Centre in Bombay and set up the new Quest Institute of LifeSciences and the Wellspring Clinical Facility in Mumbai.

Heading the task force for rapid implementation of information and technology research strategy at Nicholas Piramal, Dr. Piramal succeeded in ensuring that all manufacturing sites were ready on Y2K and were enterprise resource planning–enabled. Under her leadership, a vast wide area network using VSAT technology was put in place. She has also developed 12 cybercafes in the country for training and knowledge management of medical field representatives and doctors, started the data-warehousing solutions to manage large data bytes of information in the health care industry, and started a new project to implement Web-enabled e-business solutions.

Recognizing her specialization, the Indian government has appointed Dr. Piramal as a member of a special committee set up to transform India into a knowledge power. She has coauthored a book on nutrition and health along with Mrs. Tarla Dalal, entitled *Eat Your Way to Good Health*. She is also coauthor of another book entitled *Diet & Nutrition Guide for Patients with Renal Disease & Related Disorders*, with Dr. V. N. Acharya. She has published articles in many leading publications.

RONEN SEN

Mr. Ronen Sen assumed charge as ambassador of India to the United States in August 2004.

He began his career in the Indian Foreign Service in July 1966. From May 1968 to July 1984, he served in Indian Missions, with posts in Moscow, San Francisco, and Dhaka and in the Ministry of External Affairs, and has also been secretary to the Atomic Energy Commission of India.

From July 1984 to December 1985, Mr. Sen was joint secretary in the Ministry of External Affairs. He was thereafter joint secretary to the prime minister of India from January 1986 to July 1991, responsible for foreign affairs, defense and science & technology.

Mr. Sen was ambassador to Mexico from September 1991 to August 1992, ambassador to the Russian Federation from October 1992 to October 1998, ambassador to Germany from October 1998 to May 2002, and high commissioner to the United Kingdom from May 2002 to April 2004.

Mr. Sen participated in summit meetings in the United Nations, Commonwealth, Non-Aligned Movement, Six-Nation Five-Continent Peace Initiative, South Asian Association for Regional Cooperation, International Atomic Energy Agency, G-15, and other forums and also in over 160 bilateral summit meetings. He has had several assignments as special envoy of the prime minister of India for meetings with heads of state or governments of neighboring and other countries.

RAM SHRIRAM

Ram Shriram started Sherpalo in January 2000 with the goal of applying his wealth of operating and company-building experience to promising early-stage ventures. As a technology industry insider for over 25 years, he has worked in companies large and small across all functional areas and through fluctuating business cycles. He is always eager to roll up his sleeves and work closely with founding teams on the challenging issues that confront and sometimes confound early-stage ventures.

Mr. Shriram's knowledge of and advice on issues ranging from raising venture capital, key management hiring, making the right product choices, and defining and adapting the business model to changing market conditions has been used to secure early customer wins, build momentum from a standing start, and generate international growth.

A hallmark of his success is the ability to create the right "DNA" for a young growth company with a focus on revenue and profitability, by establishing a virtuous cycle of talented employees/owners and happy customers/users. He enjoys the process of turning founders' dreams into successful businesses and takes a long-term view that serves as a guidepost for decision making. Mr. Shriram is thoughtful, cerebral, easy to communicate with, deeply committed to the tasks he undertakes, and has a keen intuitive sense for what works in the marketplace.

Mr. Shriram has partnered with the venture capital industry and its many famous and successful members all across Silicon Valley. Because of his domain and market expertise, he brings a unique value to the building of companies with successful outcomes that fits well with the value offered by venture capital firms.

Immediately prior to founding Sherpalo, Mr. Shriram served as an officer of Amazon.com, working for Jeff Bezos, founder and CEO. He came to Amazon.com in August 1998 when Amazon acquired Junglee, an online comparison shopping firm of which Mr. Shriram was president. While at Amazon, he helped grow the customer base during its early high-growth phase in 1998–1999. Before Junglee and Amazon, Ram was a member of the Netscape executive team, joining them in 1994, before they shipped products or posted revenue. He drove the many partnerships and channels that Netscape employed to get massive distribution for its browser and server products during those now legendary early days of the Internet.

Mr. Shriram is a founding board member of Google Inc. and 247customer.com. He also serves on the boards of Plaxo, Zazzle.com, PodShow, and Business Signatures. Ram serves on the advisory board of Naukri.com, a classifieds site in India that has leading marketplaces in jobs, matrimony, and real estate serving the Indian market.

KAPIL SIBAL

Kapil Sibal is India's minister of science, technology, and ocean development. Born in Jalandhar, Punjab, on August 8, 1948, he obtained his M.A. in history from St. Stephen's College, University of Delhi, Delhi, and L.L.M. from

Harvard Law School. He joined the American Bar Association in 1972 and was designated as a senior advocate in 1983.

Prior to assuming his ministerial post in January 2006, Mr. Sibal was elected to the Indian Upper House of Parliament, the Rajya Sabha, in July 1998. In 2004, he was elected to the Lower House of the Indian Parliament, the Lok Sabha, from the historic 500-year-old Chandni Chowk constituency of Delhi.

Mr. Sibal is the leading lawyer in the Supreme Court of India and a recognized authority on constitutional law, having been involved in almost all the court's landmark cases over the past 20 years. He was elected president of the Supreme Court Bar Association in 1995 and again in 1997. He has also worked for the Human Rights Commission as a member of the Working Group on Arbitrary Detentions. Reflecting this range of experience, Mr. Sibal has contributed articles on important national and international issues such as security, nuclear proliferation, and terrorism in national dailies and periodicals.

T. S. R. SUBRAMANIAN

T. S. R. Subramanian retired from public service in India after a distinguished career spanning 37 years. He has held the highest civil service position viz the cabinet secretary to the government of India, as well as the position of chief secretary in Uttar Pradesh, the largest state of India. His other civil service posts in the government of India include secretary in the Ministry of Textiles and joint secretary to the Ministry of Commerce. He has been the agriculture production commissioner as well as the director of industries in the state government of Uttar Pradesh.

He has dealt with policy formulation and program implementation at the highest national level in India. He has been closely connected with the economic and social sectors, rural and agriculture sectors, and industry and commerce development at the federal and state levels. Significant attempts at administrative reforms were initiated by him during his tenure as cabinet secretary, including the first draft of the Right to Information Act, steps for bringing in transparency in government activities, a Citizens Charter for all public service organizations, reforms in the telecom sector, and a thrust toward improvement of the infrastructure.

For over 5 years, Mr. Subramanian worked with the International Trade Centre, a specialized trade promotion organization of the United Nations, promoted jointly by United Nations Conference on Trade and Development and World Trade Organization, as a senior adviser.

Since retirement in 1998, Mr. Subramanian has regularly contributed in the media and delivers lectures on diverse topics in many fora, including universities and other institutions. He has also authored a bestselling book *Journeys Through Babudom and Netaland—Governance in India*, a critique of governance in India; the Hindi version of the book has also been recently published. He is closely involved in the management of some large companies in India as director, and is also associated with some voluntary agencies.

Mr. Subramanian obtained his master's degree from Calcutta University, has studied at Imperial College of Science and Technology, London, and has a master's degree in public administration from Harvard University.

THOMAS WEBER

Dr. Thomas A. Weber has served at the National Science Foundation (NSF) for nearly two decades, and in February 2006 was named director of the Office of International Science and Engineering. Prior to this, Dr. Weber served for more than 10 years as the director of NSF's Materials Research Division and for 2 years as the executive officer for the Directorate for Mathematical and Physical Sciences.

In 1993, Dr. Weber was detailed to the White House and worked in the Executive Office of the President; he received the Meritorious Executive Presidential Rank Award in 1994.

Dr. Weber originally came to NSF from AT&T Bell Laboratories, where he served for 17 years as a member of the technical staff. Dr. Weber has also directed the NSF's Divisions on Advanced Scientific Computing (1988–1992) and on Information Systems (1992, 1994).

Born in Tiffin, Ohio, on June 8, 1944, Dr. Weber received his B.S. in chemistry from the University of Notre Dame in 1966. He graduated Phi Beta Kappa with a Ph.D. in chemical physics from Johns Hopkins University in 1970. Dr. Weber is an American Chemical Society fellow whose research interests include computational chemistry and materials, using computer simulation to study air pollution, polymers, glasses, liquids, metals, and semiconductor materials.

CHARLES W. WESSNER

Dr. Charles W. Wessner is a policy adviser recognized nationally and internationally for his expertise on innovation policy, including public–private partnerships, entrepreneurship, early-stage financing for new firms, and the special needs and benefits of high-technology industry. He testifies to the U.S. Congress and major national commissions, advises agencies of the U.S. government and international organizations, and lectures at major universities in the United States and abroad. Reflecting the strong global interest in innovation, he is frequently asked to address issues of shared policy interest with foreign governments, universities, and research institutes, often briefing government ministers and senior officials.

Dr. Wessner's work addresses the linkages between science-based economic growth, entrepreneurship, new technology development, university–industry clusters, regional development, small-firm finance, and public–private partnerships. His program at the National Academies also addresses policy issues associated with international technology cooperation, investment, and trade in high-technology industries. Currently, he directs a series of studies centered on government measures to encourage entrepreneurship and to support the development of new technologies. Foremost among these is a congressionally mandated study of the Small Business Innovation Research Program, reviewing the operation and achievements of this $2 billion award program for small companies and start-ups. A major review of the technology drivers of the New Economy and its sustained productivity growth is nearing completion. He is also directing a major new study on best practice in global innovation programs, entitled *Comparative Innovation Policy: Best Practice for the 21st Century*.

Appendix B

Participants List*

James Abrahamson
Stratcom

Bruce Abramson

Montek Singh Ahluwalia
Planning Commission of India

Isher Ahluwalia

Sri-Ram Aiyer
The World Bank

Sara Akbar
Oracle

Michael Alpert
Cablevision

Rajen Anand
National Federation of Indian-
   American Associations

Walter Andersen
Johns Hopkins University

Robert Armstrong
Eli Lilly and Company

Namita Arora
Metron Aviation

Sanjay Arora
Metropolitan Architects and Planners

Vasantha Arora
United News of India

Kidus Fisaha Asfaw
Department of State

George Atkinson
U.S. Department of State

Karen Autrey
National Research Council

Bruce Averill
U.S. Department of State

Sonia Baldia
Mayer, Brown, Rowe & Maw

Gautam Bambawale
Embassy of India

Geetha Bansal
National Institutes of Health

Heinrich Becker
Vistec Semiconductor Systems

Stephan Becker

Simon Bell
The World Bank

Howard Berkof
American Society of Mechanical
    Engineers

Sam Bhathena
U.S. Department of Agriculture

Yudhijit Bhattacharjee
Science

Sumant Bhutoria
Micron Technologies

John Binkley
Sandia National Laboratories

Richard Bissell
National Research Council

Peter Blair
National Research Council

William Blanpied
George Mason University

Linda Blevins
U.S. Department of Energy

Eric Bloch
Washington Advisory Group

Samuel Bodman
U.S. Department of Energy

David Brantley
U.S. Department of Commerce

Joan Burrelli
National Science Foundation

Douglas Buttrey
University of Delaware

John Cabeca
Office of the U.S. Trade
    Representative

Chuck Caprariello
Ranbaxy

Elias Carayannis
George Washington University

Melanie Carter-Maguire
Nortel

Babu Chalamala
Indocel Technologies

Krishna Challa
The World Bank

Gopi Chari

Bishwanath Chatterjee
National Institutes of Health

Anju Chaudhary
Howard University

Rohit Chaudhry
Chadbourne & Parke

Vivek Chaudhry
The World Bank

Syama Chaudhuri
University of Maryland, University
    College

Michael Cheetham
Smithsonian Institution

Chia Chen
Occupational Safety and Health
    Administration

Allen Chepuri
New Generation Advisers

Aneesh Chopra
Commonwealth of Virginia

Sid Chowdhary
Software Performance Systems

Sumitra Chowdhury
Embassy of India

M. P. Chugh
Tata AutoComp Systems

Ralph Cicerone
National Academy of Sciences

Wendy Cieslak
Sandia National Laboratories

McAlister Clabaugh
National Research Council

Kathryn Clay
Senate Energy and Natural Resources
    Committee

Kate Clemans
C&M International

Peter Cohan
Peter S. Cohan & Associates

Mildred Cooper
Luna Innovations

Ron Cooper
Small Business Administration

Melissa Coyle
C&M International

Joshua Craft
American Society of Mechanical
    Engineers

P. Y. Crawley
CNBC

Jill Dahlburg
Naval Research Laboratory

Carl J. Dahlman
Georgetown University

Ragini Dalal
The World Bank

Darlene Damm
Asia Society

Anand Das
Commerce Events

Tarun Das
Confederation of Indian Industry

Bejoy Das Gupta
Institute of International Finance

Adrian De Graaf
National Science Foundation

Ramesh Deshpande
The World Bank

Papan Devnani

Prasad Dhurjati
University of Delaware

Avinash Dikshit
Indian Science and Technology
    Ministry

Shanti Divakaran
The World Bank

Paula Dobriansky
U.S. Department of State

Tom Dogget
Reuters

Richard J. Driscoll
Office of Senator Edward Kennedy

Sisir Dutta
Howard University

Mark Dutz
The World Bank

Kamal Dwivedi
Embassy of India

Anagha Dwivedi
TV Asia

Cecile Eboa
University of the District of
    Columbia

Michael Ehst
The World Bank

Giorgio Einaudi
Embassy of Italy

Patricia Eldridge

Leland Ellis
U.S. Department of Homeland
    Security

Sarah England
Accelerating Innovation Foundation

Susan Esserman
Steptoe & Johnson

Ivy Estabrooke
National Research Council

Theela Fabian
Federal Computer Week

Irene Farkas-Conn

Ken Ferguson
U.S. Department of State

Leo Fernandes
Millenium Hospitality Consultants

Michael Fernandes
Pharma Development Holdings

Rosemary Fernandes

Guillermo Fernandez
U.S.–Mexico Foundation for Science

Kevin Finneran
Issues in Science and Technology

Susan Flack
Target

Paul Fowler
National Research Council

Merc Fox
Office of Senator Conrad

Evan Gaddis
National Electrical Manufacturers
    Association

Chris Gagne
Crowell & Moring

James Gallup
U.S. Environmental Protection
    Agency

Tanuja Garde
Office of the U.S. Trade
    Representative

Saurabh Garg
World Bank Group

Ejaz Ghani
The World Bank
Dave Ghosal
CSI Engineering

Hiten Ghosh
Hughes Network Systems

Ramya Ghosh
International Monetary Fund

Somiranjan Ghosh
Howard University

Jere Glover
Small Business Technology Council

Jamshyd Godrej
Godrej & Boyce Manufacturing

Anish Goel
U.S. Department of State

Vinod Goel
The World Bank

Anita Goel
Nanobiosym

David Goldstein
The World Bank

David Good
TATA Sons

Mary Good
University of Arkansas at Little Rock

Travis Good
America Online

Ponani S. Gopalakrishnan
International Business Machines

Ken Goretta
U.S. Air Force

Sanjay Gosain
University of Maryland

Paul Grannis
U.S. Department of Energy

Supratik Guha
International Business Machines

Alok Gupta
Excom

Anil Gupta
University of Maryland

Prasad Gupta
Advanced Technology Program

Ranjan Gupta
National Institutes of Health

Sangt Gupta
IIT

Sachin Gupta
Toucan Capital Corp.

Parvez Guzdar
University of Maryland

Diane Hanneman
National Research Council

A. Hardon
Platts

David Hart
George Mason University

Kenneth G. Herd
General Electric

Chris Hill
George Mason University

Derek Hill
National Science Foundation

Eric Holloway
U.S. Department of Commerce

Robert Holm
National Center on Education and the
    Economy

Deanna Horsley
American Association of State
    Highway Transportation Officials

Kent Hughes
Woodrow Wilson Center

John Hunter
House Committee on Government
    Reform

Jim Hurd
NanoScience Exchange

Kaye Husbands
National Science Foundation

P. V. Indiresan
Indian Institute of Technology
    (retired)

Sumathi Iyappan
University of Maryland

Murli Iyer
SAE India

Shankar Iyer
George Washington University

Sreenivasan Iyer
International Monetary Fund

Ken Jacobson
Manufacturing & Technology News

Ajay Jain
Excom

Vinod Jain
University of Maryland

S. Jaishankar
Indian Ministry of External Affairs

Phillippe Jamet
Embassy of France

Yongsuk Jang
George Washington University

R. S. Jassal
Embassy of India

Satish Jha
JM Consulting

Prakash Jha
The World Bank

Niharika Chibber Joe
Mansfield Foundation

Richard Johnson
Arnold & Porter

Regina Johnson
Platts

Rita Johnson
National Research Council

Wayne Jonas
The Samueli Institute

Ludwina Joseph
International Finance Corporation

Manoj Joshi
Embassy of India

Ianna Kachoris
Office of Senator Edward Kennedy

David Kahaner
Asian Technology Information
        Program

Greg Kalbaugh
U.S.–India Business Council

Jayant Kalotra
International Business & Technical
        Consultants, Inc.

Ajay Kalotra
International Business & Technical
        Consultants, Inc.

Chandra Kambhamettu
University of Delaware

Surinder Kapur
Sona Group

Som Karamchetty
SomeTechnologies

Flora Katz
National Institutes of Health

Pradman Kaul
Hughes Network Systems

V. Kavora
UNI

Walter Kelly
U.S. Department of State

Persis Khambatta
Asia Foundation

Anupam Khanna
The World Bank

Rita Khanna
Aeras Global Foundation

Pradeep Khosla
Carnegie Mellon University

Bharat Khurana
Uniformed Services University

Taruna Khurana
National Institutes of Health

Max Kidalov
Senate Committee on Small Business

Kartik Kilachand
World BPO Forum

Taffy Kingscott
International Business Machines

Mike Kirk
American Intellectual Property Law
    Association

Mindy Kotler
Asia Policy Point

Jag Kottha
ASEI

Srivatsa Krishna
The World Bank

Sanjee Krishnan
The World Bank

Harish Krishnan
International Business Machines

Satish Kulkarni
University of California

Amar Kumar
AAA IT & Management Consultants

Dhanendra Kumar
The World Bank

Manmohan Kumar
International Monetary Fund

Mukesh Kumar
Technical Resources International

Satyendra Kumar
National Science Foundation

Vinay Kumar
University of Chicago

Virender Kumar
Embassy of India

Y. P. Kumar
Indian Science and Technology
    Ministry

Vivek Kundra
Commonwealth of Virginia

Anne Kuriakose
The World Bank

Martha LaCrosse
BAE Systems Inc.

Keren Ladin
Institute of Medicine

Erin Lamos
National Research Council

Jeff Lande
Information Technology Association
    of America

Christina Landi
U.S. Department of State

Noel Le
Progress & Freedom Foundation

Burton Lee
National Research Council

Susan Lee
Center for American Progress

Rolf Lehming
National Science Foundation

Heather LeMunyon
Accelerating Innovation Foundation

Jamie Link
Office of Senator Lieberman

Nigel Lockyer
University of Pennsylvania

Marjorie Lueck
National Science Foundation

Elizabeth Lyons
National Science Foundation

Will Maher
Ernst & Young

Nabanjan Maitra
Henry Stimson Center

Promodh Malhotra
Global Finance Associates

Vilas Mandlekar
The World Bank

Dev Mani
The National Academies

Suresh Maniam
Mayer Brown Rowe & Mawe

John Marburger
White House Office of Science &
    Technology Policy

R. A. Mashelkar
Council on Scientific and Industrial
    Research

Paul Matthew
EPMG Consulting

Bill McCluskey
U.S. Department of Defense

David McCormick
U.S. Department of Commerce

Mac McCullough
National Research Council

Neil McDonald
Federal Technology Watch

Diane McMahon
Stonebridge International

S. Ahmed Meer
ICF International

Natalia Melcer
National Research Council

Stephen A. Merrill
National Research Council

Egils Milbergs
Center for Accelerating
  Innovation

Jeffrey Miotke
U.S. Department of State

Abhay Nath Mishra
University of Maryland

Shekhar Mishra
Fermi National Laboratory

Lakshmi Mishra
Ayurvedic Health Care & Toxicology
  Center

Anoop Mishra
Embassy of India

B. P. Misra
International Monetary Fund

Nyema Mitchell
Institute for Defense Analysis

Raja Mitra
The World Bank

Amit Mittal

Sushanta Mohapatra
SRI International

Andrew Molchany
American Chemical Society

David Morgenthaler
Morgenthaler

Francisco Moris
National Science Foundation

Bhumika Muchhala
Woodrow Wilson Center

Robert Muir
Tata Ryerson

Rob Mulligan
American Electronics Association

Prabhat Munshi
Indian Institute of Technology

Kazue Muroi
Washington Core

Helen Murray
American Association of State
    Highway Transportation Officials

Shalizeh Nadjmi
U.S. Department of Commerce

Sarita Nagpal
Confederation of Indian Industry

Anita Nahal
Howard University

Sreekumar Nair
Confederation of Indian Industry

Arvind Nandelkar
Howard University

Anant Narayanan
InnovaLex

Kesh Narayanan
National Science Foundation

Sara Nerlove
National Science Foundation

Harvey Newman
California Institute of Technology

Ruchika Nijhara
University of Maryland

Paul op den Brouw
Embassy of the Royal Netherlands

Mary O'Driscoll
Greenwire

Daniel O'Neill
FasterCures

Inja Paik
U.S. Department of Energy

Diane Palmintera
Innovation Associates

Guru Parulkar
National Science Foundation

Amir Pasic
George Washington University

Kiran Pasricha
Confederation of Indian Industry

B. Patel
Adept

Praful Patel
The World Bank

Vainav Patel
National Institutes of Health

Ashok Patil
U.S. Army

Mike Patterson
WorldTech

Dilip Paul
ACES, Inc.

Kristin Paulson
United Technologies

Paul Pavwoski
U.S. Department of Commerce

Eija Pehu
The World Bank

Frank Pennypacker
Office of the Secretary of Defense

Pierre Perrolle
National Science Foundation

Nicolas Peter
George Washington University

P. Kim Pham
National Institutes of Health

Sanjay Phogat
National Institutes of Health

Swati Piramal
Nicholas Piramal India Ltd.

Linda Powers
Toucan Capital Corp.

Nitin Pradhan
Fairfax County Public Schools

Ernest Preeg
Manufacturers Alliance

Surya Raghu
Advanced Fluidics

Gulshan Rai
Department of Energy

Juhi Raikhy
The Moksha Group

Neerja Raman
Stanford University

Ardhanari Ramaswamy
International Finance Corporation

Priya Ramesh
Luna Innovations

Ranga Rangarajan
WorldSpace

Maya Randall
Dow Jones

Mukul Ranjan
National Institutes of Health

Alan Rapoport
National Science Foundation

Pradeep Rau
George Washington University

Pramud Rawat
AIAA

Carl Ray
Natuonal Aeronautics and Space
 Administration

Devasish Ray
TV Asia

Vivian Reed
Caterpillar

Jay Rengarajan
The World Bank

Andy Reynolds
U.S. Department of State

Volker Rieke
Embassy of Germany

Thyra Riley
The World Bank

Ben Roberts
National Research Council

Joan Rolf
U.S. Office of Science & Technology
 Policy

Dan Roos
Massachusetts Institute of
 Technology

Narendra Rustagi
Howard University

Krishnan Sabnani
Lucent Technologies

Jerald Sadoff
Aeras Global Foundation

Nishit Sahay
ANTHRA Technologies

Leila Saldanha
NUTRIQ

Ishaq Saleem
Delaware Economic Development
   Office

Bill Salmon
Council of Academies of Engineering
   and Technological Sciences

R. C. Saravanabhavan
Howard University

Anuja Saurkar
International Food Policy Research
   Institute

Ravi Sawhney
National Institutes of Health

Arshad Sayed
The World Bank

Vijay Sazawal
USEC, Inc.

Ted Schmitt
National Research Council

Joel Schnur
Naval Research Laboratory

Jack Schofield
Sensoars Electronics

Ken Schramko
SEMI

Danny Sebright
The Cohen Group

Chandra Sekaran
The World Bank

Ronen Sen
Embassy of India

Debdatta Sengupta
International Food Policy Research
   Institute

Shiladitya Sengupta
Massachusetts Institute of Technology

Arun Seraphin
Senate Armed Services Committee

Veronica Seva Gonzalez
Brandon Green & Associates

Kapil Sharma
Tata Group

Ravi Sharma
General Motors

Prabha Sharma

Faruk Sheikh
Food and Drug Administration

Sudhakar Shenoy
Information Management Consultants

Nathan Shepherd
U.S. Department of State

Hideo Shindo
NEDO

Stephanie Shipp
Advanced Technology Program

Jayasankar Shivakumar
The World Bank

Sujai Shivakumar
National Research Council

Debashish Shome
Neva Group

Ram Shriram
Google

Mark Shuart
Office of Congressman Lewis

J. Shukla
George Mason University

Kapil Sibal
Indian Ministry of Science and
*   Technology*

Promila Sibal

Rob Sienkiewicz
Advanced Technology Program

J. P. Singh
George Washington University

Nathan Singh
U.S. Department of State

Ritin Singh
The World Bank

Alok Sinha
Penn State

Shrikant Sinha
Adnet

Shankar Sitapati
American Red Cross

Dakshanamurthy Sivanesan
Georgetown University

Siddharth Sivaraman
Stratcom

Amanda Slocum
National Science Foundation

Joseph Snyder
Asia Society

Michael Snyder
American Chemical Society

Anne Solomon
CSIS

Ron Somers
U.S.–India Business Council

Anuja Sonalker
Sparta

Shobhana Sosale
The World Bank

Anil Srivastava
Capital Technology Information
    Services

Robin Staffin
U.S. Department of Energy

Marc Stanley
Advanced Technology Program

T. S. R. Subramanian
Government of India (retired)

Istvan Takacs
Embassy of Hungary

Andrew Tein
C&M International

Vasant Teland
Howard University

Paul Tennassee
University of the District of Columbia

Nitish Thakor
Johns Hopkins University

Fabio Thiers
Massachusetts Institute of Technology

Brajendra Tripathi
National Institutes of Health

Brian Tsou
U.S. Department of Commerce

Jim Turner
House Committee on Science

Anuja Utz
The World Bank

Chandrasekhar Vallath
PI-Tech

Dhruv Varma
Vasucorp

Garrett Vaughn
Manufacturers Alliance

Venky Venkatesan
Neocera

Anubha Verma
Georgetown University

Ray Vickery
Stonebridge International

Darshana Vyas
Counterpart International

Lilya Wagner
Counterpart International

Thomas A. Weber
National Science Foundation

Marjorie Weisskohl
U.S. Department of Commerce

Charles W. Wessner
National Research Council

Jody Westby
Global Cyber Risk

Jeff Williams
Civilian R&D Foundation

Joan Winston
National Research Council

Patricia Wrightson
National Research Council

Amanda Yarnell
Chemical and Engineering News

Ilyse Zable
International Finance Corporation

Appendix C

Bibliography

Aghion, Phillipe, Robin Burgess, Stephen Redding, and Fabrizio Zilibotti. 2003. "The Unequal Effects of Liberalization: Theory and Evidence from India." Center for Economic Policy Research, March.

Ahluwalia, Montek Singh. 2001. "State Level Reforms Under Economic Reforms in India." Stanford University Working Paper No. 96, March.

Ahluwalia, Montek Singh. 2002. "Economic Reforms in India Since 1991: Has Gradualism Worked?" *Journal of Economic Perspectives* 16(2):67–88.

Alic, John A., Lewis M. Branscomb, Harvey Brooks, Ashton B. Carter, and Gerald L. Epstein. 1992. *Beyond Spin-off: Military and Commercial Technologies in a Changing World*. Boston, MA: Harvard Business School Press.

Amsden, Alice H. 2001. *The Rise of "the Rest": Challenges to the West from Late-industrializing Economies*. Oxford, UK: Oxford University Press.

Amsden, Alice H., and Wan-wen Chu. 2003. *Beyond Late Development: Taiwan's Upgrading Policies*. Cambridge, MA: MIT Press.

Amsden, Alice H., Ted Tschang, and Akira Goto. 2001. "Do Foreign Companies Conduct R&D in Developing Countries?" Tokyo, Japan: ADB Institute.

Annamalai, Kuttayan, and Sachin Rao. 2003. "What Works: ITC's E-Choupal and Profitable Rural Transformation." World Resources Institute. Accessed at <*http://www.digitaldividend.org/case/case_echoupal.htm*>.

Archibugi, Danielle, Jeremy Howells, and Jonathan Michie, eds. 1999. *Innovation Policy and the Global Economy*. Cambridge, UK: Cambridge University Press.

Asheim, Bjorn T., Arne Isaksen, Claire Nauwelaers, and Franz Todtling, eds. 2003. *Regional Innovation Policy for Small-Medium Enterprises*. Cheltenham, UK, and Northampton, MA: Edward Elgar.

Associated Press. 2006. "Indian Drug Maker Nicholas Piramal in Deal to Acquire Pfizer's UK Facility." June 15.

Associated Press. 2006. "India Shows Confidence, Openness About Risks Confronting Economy." November 28.

Athreye, Suma S. 2000. "Technology Policy and Innovation: The Role of Competition Between Firms." In Pedro Conceicao, Syed Shariq, and Manuel Heitor, eds. *Science, Technology, and Innovation Policy: Opportunities and Challenges for the Knowledge Economy*. Westport, CT and London, UK: Quorum Books.

Audretsch, David B. 2006. *The Entrepreneurial Society*, Oxford, UK: Oxford University Press.

Audretsch, David B., Heike Grimm, and Charles W. Wessner. 2005. *Local Heroes in the Global Village: Globalization and the New Entrepreneurship Policies*. New York: Springer.

Audretsch, David B., ed. 1998. *Industrial Policy and Competitive Advantage*, Volumes 1 and 2. Cheltenham, UK: Edward Elgar.

Audretsch, D. B., B. Bozeman, K. L. Combs, M. P. Feldman, A. N. Link, D. S. Siegel, P. Stephan, G. Tassey, and C. Wessner. 2002. "The Economics of Science and Technology." *Journal of Technology Transfer* 27:155–203.

Baddeley, Michelle, Kirsty McNay, and Robert Cassen. 2006. "Divergence in India: Income Differentials at the State Level, 1970–97." *Journal of Development Studies* 42(6):1000–1022.

Baldwin, John Russel, and Peter Hanel. 2003. *Innovation and Knowledge Creation in an Open Economy: Canadian Industry and International Implications*. Cambridge, UK: Cambridge University Press.

Balzat, Markus, and Andreas Pyka. 2006. "Mapping National Innovation Systems in the OECD Area." *International Journal of Technology and Globalisation* 2(1–2):158–176.

Bartzokas, Anthony, and Morris Teubal. 2002. "The Political Economy of Innovation Policy Implementation in Developing Countries." *Economics of Innovation and New Technology* 11(4–5).

Bhidé, Amar. 2006. "Venturesome Consumption, Innovation and Globalization." Paper presented at the Centre on Capitalism & Society and CESifo Venice Summer Institute 2006, "Perspectives on the Performance of the Continent's Economies," Venice International University. San Servolo, Italy, July 21–22.

Biegelbauer, Peter S., and Susana Borras, eds. 2003. *Innovation Policies in Europe and the U.S.: The New Agenda*. Aldershot, UK: Ashgate.

Blomström, Magnus, Ari Kokko, and Fredrik Sjöholm. 2002. "Growth & Innovation Policies for a Knowledge Economy: Experiences from Finland, Sweden, & Singapore." EIJS Working Paper, Series No. 156.

Bloomberg News. 2006. "The Next Green Revolution." August 21.

Borras, Susana. 2003. *The Innovation Policy of the European Union: From Government to Governance*. Cheltenham, UK: Edward Elgar.

Borrus, Michael, and Jay Stowsky. 2000. "Technology Policy and Economic Growth." In Charles Edquist and Maureen McKelvey, eds. *Systems of Innovation: Growth, Competitiveness and Employment*, Vol. 2. Cheltenham, UK and Northampton, MA: Edward Elgar.

Branscomb, Lewis M., and Philip E. Auerswald. 2002. *Between Invention and Innovation: An Analysis of Funding for Early-Stage Technology Development*. NIST GCR 02–841. Gathersburg, MD: National Institute of Standards and Technology. November.

Buchanan, James M. 1987. "An Economic Theory of Clubs." In *Economics: Between a Predictive Science and Moral Philosophy*. College Station, TX: Texan A&M University Press, 1987.

Caracostas, Paraskevas, and Ugur Muldur. 2001. "The Emergence of the New European Union Research and Innovation Policy." In P. Laredo and P. Mustar, eds. *Research and Innovation Policies in the New Global Economy: An International Comparative Analysis*. Cheltenham, UK: Edward Elgar.

Chand, Satish, and Kunal Sen. 2002. "Trade Liberalization and Productivity Growth: Evidence from Indian Manufacturing." *Review of Development Economics* 6, February.

Chesbrough, Henry. 2003. *Open Innovation: The New Imperative for Creating and Profiting from Technology*. Cambridge, MA: Harvard Business School Press.

Cimoli, Mario, and Marina della Giusta. 2000. "The Nature of Technological Change and Its Main Implications on National and Local Systems of Innovation." IIASA Interim Report IR-98-029.

Coburn, Christopher, and Dan Berglund. 1995. *Partnerships: A Compendium of State and Federal Cooperative Programs.* Columbus, OH: Battelle Press.

Combs, Kathryn L., and Albert N. Link. 2003. "Innovation Policy in Search of an Economic Paradigm: The Case of Research Partnerships in the United States." *Technology Analysis & Strategic Management* 15(2).

Council on Competitiveness. 2005. *Innovate America: Thriving in a World of Challenge and Change.* Washington, D.C.: Council on Competitiveness.

Dahlman, Carl J., and Jean Eric Aubert. 2001. *China and the Knowledge Economy: Seizing the 21st Century.* Washington, D.C.: World Bank.

Dahlman, Carl, and Anuja Utz. 2005. *India and the Knowledge Economy: Leveraging Strengths and Opportunities.* Washington, D.C.: World Bank.

Daneke, Gregory A. 1998. "Beyond Schumpeter: Non-linear Economics and the Evolution of the U.S. Innovation System." *Journal of Socio-economics* 27(1):97–117.

Das, Gurcharan. 2006. "The India Model." *Foreign Affairs* 85(4).

De la Mothe, John, and Gilles Paquet. 1998. "National Innovation Systems, 'Real Economies' and Instituted Processes." *Small Business Economics* 11:101–111.

Doloreux, David. 2004. "Regional Innovation Systems in Canada: A Comparative Study." *Regional Studies* 38(5):479–492.

Domain-b. 2005. "Nasscom-McKinsey: India to Face Skilled Workers' Shortage by Next Decade," Dec. 12, 2005. Access at <*http://www.domainb.com/organisation/nasscom/20051217_shortage.html*>.

Eaton, Jonathan, Eva Gutierrez, and Samuel Kortum. 1998. "European Technology Policy." NBER Working Paper 6827.

The Economist. 2006. "India's Acquisition Spree." October 12.

The Economist. 2006. "India's Special Economic Zones." October 12.

Edler, J., and S. Kuhlmann. 2005. "Towards One System? The European Research Area Initiative, the Integration of Research Systems and the Changing Leeway of National Policies." *Technikfolgenabschätzung: Theorie und Praxis.* 1(4):59–68.

Eickelpasch, Alexander, and Michael Fritsch. 2005. "Contests for Cooperation: A New Approach in German Innovation Policy." *Research Policy* 34:1269–1282.

Endquist, Charles, ed. 1997. *Systems of Innovation: Technologies, Institutions, and Organizations,* London, UK: Pinter.

EOS Gallup Europe. 2004. *Entrepreneurship.* Flash Eurobarometer 146. January. Accessed at <*http://ec.europa.eu/enterprise/enterprise_policy/survey/eurobarometer146_en.pdf*>.

European Commission. 2003. "Innovation in Candidate Countries: Strengthening Industrial Performance." May.

Fangerberg, Jan. 2002. *Technology, Growth, and Competitiveness: Selected Essays.* Cheltenham, UK, and Northampton, MA: Edward Elgar.

Federal Register Notice. 2004. "2004 WTO Dispute Settlement Proceeding Regarding China: Value-Added Tax on Integrated Circuits." April 21.

Feldman, Maryann, and Albert N. Link. 2001. "Innovation Policy in the Knowledge-Based Economy." In *Economics of Science, Technology and Innovation,* Vol. 23. Boston, MA: Kluwer Academic Press.

Feldman, Maryann P., Albert N. Link, and Donald S. Siegel. 2002. *The Economics of Science and Technology: An Overview of Initiatives to Foster Innovation, Entrepreneurship, and Economic Growth.* Boston, MA: Kluwer Academic Press.

The Financial Times. 2003. "India's Islands of Excellence Under Pressure." February 21.

The Financial Times. 2006. "India Needs Big Infrastructure Drive." February 23.

The Financial Times. 2006. "Engaging India: Demographic Dividend or Disaster?" November 15.

Fonfria, Antonio, Carlos Diaz de la Guardia, and Isabel Alvarez. 2002. "The Role of Technology and Competitiveness Policies: A Technology Gap Approach." *Journal of Interdisciplinary Economics* 13:223–241.

Foray, Dominique, and Patrick Llerena. 1996. "Information Structure and Coordination in Technology Policy: A Theoretical Model and Two Case Studies," *Journal of Evolutionary Economics* 6(2):157–173.

Friedman, Thomas. 2005. *The World Is Flat: A Brief History of the 21st Century*. New York: W. H. Freeman.

Furman, Jeffrey L., Michael E. Porter, and Scott Stern. 2002. "The Determinants of National Innovative Capacity." *Research Policy* 31:899–933.

George, Gerard, and Ganesh N. Prabhu. 2003. "Developmental Financial Institutions as Technology Policy Instruments: Implications for Innovation and Entrepreneurship in Emerging Economies." *Research Policy* 32(1):89–108.

Giridharadas, Anand. 2006. "Growth Spurt in India Hides Government Gridlock," *International Herald Tribune*, September 29.

Giridharadas, Anand. 2006. "In India's Higher Education, Few Prizes for 2nd Place." *International Herald Tribune*. November 16.

Grande, Edgar. 2001. "The Erosion of State Capacity and European Innovation Policy: A Comparison of German and EU Information Technology Policies." *Research Policy* 30(6):905–921.

Hall, Bronwyn H. 2002. "The Assessment: Technology Policy." *Oxford Review of Economic Policy* 18(1):1–9.

The Hindu. 2005. "State Pins Hope on Growth Initiative." August 4.

The Hindu. 2006. "Stress on Innovation in Manufacturing." November 17.

Huang, Yasheng, and Tarun Khanna. 2003. "Can India Overtake China?" *Foreign Policy* July–August.

Hughes, Kent. 2005. *Building the Next American Century: The Past and Future of American Economic Competitiveness*. Washington, D.C.: Woodrow Wilson Center Press. Chapter 14.

Hughes, Kent H. 2005. "Facing the Global Competitiveness Challenge." *Issues in Science and Technology* XXI(4):72–78.

International Herald Tribune. 2006. "Start-Ups Explore Abroad for IPOs." December 25.

Jaffe, Adam B, Josh Lerner, and Scott Stern, eds. 2003. *Innovation Policy and the Economy,* Vol. 3. Cambridge, MA: MIT Press.

Jasanoff, Sheila, ed. 1997. *Comparative Science and Technology Policy*. Elgar Reference Collection. International Library of Comparative Pubic Policy, Vol. 5. Cheltenham, UK and Lyme, NH: Edward Elgar.

Joy, William. 2000. "Why the Future Does Not Need Us." *Wired* 8(April).

Kapur, Devesh. 2003. "Indian Diaspora as a Strategic Asset." *Economic and Political Weekly* 38(5): 445–448.

Kapur, Surinder. 2006. "Nurture New Technology and Innovation, Stay Competitive." *The Financial Express*. November 16.

Koschatzky, Knut. 2003. "The Regionalization of Innovation Policy: New Options for Regional Change?" In G. Fuchs and Phil Shapira, eds. *Rethinking Regional Innovation: Path Dependency or Regional Breakthrough?* London, UK: Kluwer.

Kuhlmann, Stephan, and Jakob Edler. 2003. "Scenarios of Technology and Innovation Policies in Europe: Investigating Future Governance—Group of 3." *Technological Forecasting & Social Change* 70.

Kumar, Deepak. 1995. *Science and the Raj: 1857–1905*. New Delhi and New York: Oxford University Press.

Lall, Sanjaya. 2002. "Linking FDI and Technology Development for Capacity Building and Strategic Competitiveness." *Transnational Corporations* 11(3):39–88.

Laredo, Philippe, and Philippe Mustar, eds. 2001. *Research and Innovation Policies in the New Global Economy: An International Perspective*. Cheltenham, UK: Edward Elgar.

Lembke, Johan. 2002. *Competition for Technological Leadership: EU Policy for High Technology*. Cheltenham, UK, and Northampton, MA: Edward Elgar.

Lemola, Tarmo. 2002. "Convergence of National Science and Technology Policies: The Case of Finland." *Research Policy* 31(8–9):1481–1490.

Leslie, Stuart, and Robert Kargon. 2006. "Exporting MIT." *Osiris* 21:110–130.

Lewis, James A. 2005. *Waiting for Sputnik: Basic Research and Strategic Competition*. Washington, D.C.: Center for Strategic and International Studies.

Lin, Otto. 1998. "Science and Technology Policy and Its Influence on the Economic Development of Taiwan." In Henry S. Rowen, ed. *Behind East Asian Growth: The Political and Social Foundations of Prosperity*. London, UK, and New York: Routledge.

Maddison, Angus, and Donald Johnston. 2001. *The World Economy: A Millennial Perspective*. Paris, France: Organisation for Economic Co-operation and Development.

Mani, Sunil. 2004. "Government, Innovation and Technology Policy: An International Comparative Analysis." *International Journal of Technology and Globalization* 1(1).

McKibben, William. 2003. *Enough: Staying Human in an Engineered Age*. New York, NY: Henry Holt & Co.

Meyer-Krahmer, Frieder. 2001. Industrial Innovation and Sustainability—Conflicts and Coherence." Pp. 177–195 in Daniele Archibugi, Bengt–Ake Lundvall, eds. *The Globalizing Learning Economy*. New York: Oxford University Press.

Meyer-Krahmer, Frieder. 2001. "The German Innovation System." Pp. 205–252 in P. Larédo and P. Mustar, eds. *Research and Innovation Policies in the New Global Economy: An International Comparative Analysis*. Cheltenham, UK: Edward Elgar.

Mitra, Raja. 2006. "India's Potential as a Global R&D Power." In Magnus Karlsson, ed. *The Internationalization of Corporate R&D*. Östersund: Swedish Institute for Growth Policy Studies.

Mody, Ashok. and Carl Dahlman. 1992. "Performance and Potential of Information Technology: An International Perspective." *World Development* 20(12).

Murali, Kanta. 2003. "The IIT Story: Issues and Concerns." *Frontline,* 20(3). Accessed at <*http://www.flonnet.com/fl2003/stories/20030214007506500.htm*>.

Mustar, Phillipe. and Phillipe Laredo. 2002. "Innovation and Research Policy in France (1980–2000) or The Disappearance of the Colbertist State." *Research Policy* 31:55–72.

National Academy of Engineering. 2004. *The Engineer of 2020: Visions of Engineering in the New Century*. Washington, D.C.: The National Academies Press.

National Academy of Sciences, National Academy of Engineering, and Institute of Medicine. 2007. *Rising Above the Gathering Storm: Energizing and Employing America for a Brighter Economic Future*. Washington, D.C.: National Academies Press.

National Manufacturing Competitiveness Council. 2006. "The National Strategy for Manufacturing," New Delhi, March.

National Research Council. 1996. *Conflict and Cooperation in National Competition for High-Technology Industry*. Washington, D.C.: National Academy Press.

National Research Council. 1999. *The Advanced Technology Program: Challenges and Opportunities*. Charles W. Wessner, ed. Washington, D.C.: National Academy Press.

National Research Council. 1999. *Funding a Revolution: Government Support for Computing Research*. Washington, D.C.: National Academy Press.

National Research Council. 1999. *Industry-Laboratory Partnerships: A Review of the Sandia Science and Technology Park Initiative*. Charles W. Wessner, ed. Washington, D.C.: National Academy Press.

National Research Council. 1999. *New Vistas in Transatlantic Science and Technology Cooperation*. Charles W. Wessner, ed. Washington, D.C.: National Academy Press.

National Research Council. 1999. *The Small Business Innovation Research Program: Challenges and Opportunities*. Charles W. Wessner, ed. Washington, D.C.: National Academy Press.

National Research Council. 1999. *U.S. Industry in 2000: Studies in Competitive Performance*. David C. Mowery, ed. Washington, D.C.: National Academy Press.

National Research Council. 2000. *The Small Business Innovation Research Program: A Review of the Department of Defense Fast Track Initiative.* Charles W. Wessner, ed. Washington, D.C.: National Academy Press.

National Research Council. 2001. *The Advanced Technology Program: Assessing Outcomes.* Charles W. Wessner, ed. Washington, D.C.: National Academy Press.

National Research Council. 2001. *Building a Workforce for the Information Economy.* Washington, D.C.: National Academy Press.

National Research Council. 2001. *Capitalizing on New Needs and New Opportunities: Government-Industry Partnerships in Biotechnology and Information Technologies.* Charles W. Wessner, ed. Washington, D.C.: National Academy Press.

National Research Council. 2001. *A Review of the New Initiatives at the NASA Ames Research Center.* Charles W. Wessner, ed. Washington, D.C.: National Academy Press.

National Research Council. 2001. *Trends in Federal Support of Research and Graduate Education.* Stephen A. Merrill, ed. Washington, D.C.: National Academy Press.

National Research Council. 2003. *Government-Industry Partnerships for the Development of New Technologies: Summary Report.* Charles W. Wessner, ed. Washington, D.C.: The National Academies Press.

National Research Council. 2004. *The Small Business Innovation Research Program: Program Diversity and Assessment Challenges.* Charles W. Wessner, ed. Washington, D.C.: The National Academies Press.

National Research Council. 2005. *Getting Up to Speed: The Future of Superconducting*, Susan L. Graham, Marc Snir, and Cynthia A. Patterson, eds. Washington, D.C.: The National Academies Press.

National Research Council. 2007. *Enhancing Productivity Growth in the Information Age: Measuring and Sustaining the New Economy.* Dale W. Jorgenson and Charles W. Wessner, eds. Washington, D.C.: The National Academies Press.

National Research Council. 2007. *SBIR and the Phase III Challenge of Commercialization.* Charles W. Wessner, ed. Washington, D.C.: The National Academies Press.

National Research Council. 2007. *Innovation Policies for the 21st Century.* Charles W. Wessner, ed. Washington, D.C.: The National Academies Press.

Nelson, Richard R., and Katherine Nelson. 2002. "Technology, Institutions, and Innovation Systems." *Research Policy* 31:265–272.

Nelson, Richard R., and Nathan Rosenberg. 1993. "Technical Innovation and National Systems." In Richard R. Nelson, ed. *National Innovation Systems: A Comparative Analysis.* Oxford, UK: Oxford University Press.

Organisation for Economic Co-operation and Development. 1999. *Boosting Innovation: The Cluster Approach.* Paris, France: Organisation for Economic Co-operation and Development.

Organisation for Economic Co-operation and Development. 1999. *Managing National Innovation Systems,* Paris, France and Washington, D.C.: Organisation for Economic Co-operation and Development.

Organisation for Economic Co-operation and Development. 2001. *Social Sciences and Innovation,* Washington, D.C.: Organisation for Economic Co-operation and Development.

Organisation for Economic Co-operation and Development. 2004. "Summary Report: Micro-policies for Growth and Productivity." DSTI/IND(2004)7 Paris, France: Organisation for Economic Co-operation and Development. October.

Oughton, Christine. 1997. "Competitiveness in the 1990s." *Economic Journal* 107(444):1486–1503.

Oughton, Christine, Mikel Landabaso, and Kevin Morgan. 2002. "The Regional Innovation Paradox: Innovation Policy and Industrial Policy." *Journal of Technology Transfer* 27(1).

Patel, P., and K. Pavitt. 1994. "National Innovation Systems: Why They Are Important and How They Might Be Compared?" *Economic Change and Industrial Innovation.*

Posen, Adam S. 2001. "Japan." In Benn Steil, David G. Victor, and Richard R. Nelson, eds. *Technological Innovation and Economic Performance*. Princeton, NJ: Princeton University Press.

President's Council of Advisors on Science and Technology. 2004. "Sustaining the Nation's Innovation System: Report on Information Technology Manufacturing and Competitiveness." Washington, D.C.: Executive Office of the President. January.

The Press Trust of India. 2006. "Government Preparing to Reply to Supreme Court on Quotas." June 16.

PricewaterhouseCoopers. 2006. "China's Impact on the Semiconductor Industry: 2005 Update." PricewaterhouseCoopers.

Rai, Saritha. 2006. "India Becoming a Crucial Cog in the Machine at I.B.M." *New York Times*. June 5.

Reuters. 2006. "China Sees No Quick End to Economic Boom." February 21.

Rodrik, Dani, and Arvind Subramanian. 2004. "From 'Hindu Growth' to Productivity Surge: The Mystery of the Indian Growth Transition," NBER Working Paper 10376.

Romanainen, Jari. 2001. "The Cluster Approach in Finnish Technology Policy." Pp. 377–388 in Edward M. Bergman, Pim den Hertog, and David Charles, eds. *Innovative Clusters: Drivers of National Innovation Systems*. OECD Proceedings. Washington, D.C.: Organisation for Economic Co-operation and Development.

Ruttan, Vernon. 2002. *Technology, Growth and Development: An Induced Innovation Perspective*. Oxford, UK: Oxford University Press.

Rutten, Roel, and Frans Boekema. 2005. "Innovation, Policy and Economic Growth: Theory and Cases." *European Planning Studies* 13(8).

Schaffer, Teresita. 2002. "Building a New Partnership with India." *Washington Quarterly* 25(2): 31–44.

Scherer, F. M. 2001. "U.S. Government Programs to Advance Technology." *Revue d'Economie Industrielle* 0(94):69–88.

Sheehan, Jerry, and Andrew Wyckoff. 2003. "Targeting R&D: Economic and Policy Implications of Increasing R&D Spending." DSTI/DOC(2003)8. Paris, France: Organisation for Economic Co-operation and Development.

Shin, Roy W. 1997. "Interactions of Science and Technology Policies in Creating a Competitive Industry: Korea's Electronics Industry." *Global Economic Review* 26(4):3–19.

Singh, Manmohan. 2006. Remarks by the Chairman of the Planning Commission on the release of the 11th Plan Approach Paper. Accessed at *<http://pmindia.nic.in/speech/content.asp?id=431>*. October 18

Smits, Ruud, and Stefan Kuhlmann. 2004. "The Rise of Systemic Instruments in Innovation Policy." *International Journal of Foresight and Innovation Policy*. 1(1/2).

Soete, Luc G., and Bastiann J. ter Weel. 1999. "Innovation, Knowledge Creation and Technology Policy: The Case of the Netherlands." *De Economist* 147(3).

Stanford University. 1999. *Inventions, Patents and Licensing: Research Policy Handbook*. Document 5.1. July 15.

Talele, Chitram J. 2003. "Science and Technology Policy in Germany, India and Pakistan." *Indian Journal of Economics and Business* 2(1):87–100.

Tansley, A. G. 1939. "British Ecology During the Past Quarter Century: The Plant Community and the Ecosystem." *Journal of Ecology* 27(2):513–530.

Tassey, Gregory. 2004. "Policy Issues for R&D Investment in a Knowledge-Based Economy." *Journal of Technology Transfer* 29:153–185.

Teubal, Morris. 2002. "What is the Systems Perspective to Innovation and Technology Policy and How Can We Apply It to Developing and Newly Industrialized Economies?" *Journal of Evolutionary Economics* 12(1–2).

U.S. Department of Energy. 2006. Press Release. "Department Requests $4.1 Billion Investment as Part of the American Competitiveness Initiative: Funding to Support Basic Scientific Research." February 2.

U.S. General Accounting Office. 2002. *Export Controls: Rapid Advances in China's Semiconductor Industry Underscore need for Fundamental U.S. Policy Review.* GAO-020620. Washington, D.C.: U.S. General Accounting Office. April.

Wade, Robert. 1985. "The Market for Public Office: Why the Indian State Is not Better at Development." *World Development* 13:467–497.

Washington Post. 2006. "Chinese to Develop Sciences, Technology." February 10, p. A16

Wessner, Charles W. 2005. "Entrepreneurship and the Innovation Ecosystem." In David B. Audretsch, Heike Grimm, and Charles W. Wessner, eds. *Local Heroes in the Global Village: Globalization and the New Entrepreneurship Policies.* New York, NY: Springer.

White House. 2006. "Fact Sheet: United States and India: Strategic Partnership." Press Release. March 2.

World Bank. 2004. *Innovation Systems: World Bank Support of Science and Technology Development.* Vinod Kumar Goel, ed. Washington, D.C.: World Bank.

World Bank. 2006. *The Environment for Innovation in India.* South Asia Private Sector Development and Finance Unit. Washington, D.C.: World Bank.

World Bank. 2006. *World Development Indicators.* Accessed at *<http://devdata.worldbank.org/wdi2006/contents/cover.htm>*.

World Bank International Finance Corporation. 2006. *Doing Business in 2006: Creating Jobs.* Washington, D.C.: International Bank for Reconstruction and Development.

Yukio, Sato. 2001. "The Structure and Perspective of Science and Technology Policy in Japan." In Phillipe Laredo and Phillipe Mustar, eds. *Research and Innovation Policies in the New Global Economy: An International Comparative Analysis.* Cheltenham, UK and Northampton, MA: Edward Elgar.

Zeigler, Nicholas J. 1997. *Governing Ideas: Strategies for Innovation in France and Germany.* Ithaca, NY, and London, UK: Cornell University Press.